Assignments in
Applied Statistics

Assignments in Applied Statistics

edited by
Simon Conrad

School of Management
University of Manchester Institute of Science and Technology

JOHN WILEY & SONS

Chichester · New York · Brisbane · Toronto · Singapore

Wiley Editorial Offices

John Wiley & Sons Ltd, Baffins Lane, Chichester,
West Sussex PO19 1UD, England

John Wiley & Sons, Inc., 605 Third Avenue,
New York, NY 10158–0012, USA

Jacaranda Wiley Ltd, G.P.O. Box 859, Brisbane,
Queensland 4001, Australia

John Wiley & Sons (Canada) Ltd, 22 Worcester Road,
Rexdale, Ontario M9W 1L1, Canada

John Wiley & Sons (SEA) Pte Ltd, 37 Jalan Pemimpin #05–04,
Block B, Union Industrial Building, Singapore 2057

Library of Congress Cataloging-in-Publication Data:
Assignments in applied statistics/edited by Simon Conrad.
    p.   cm.
  Bibliography: p.
  ISBN 0 471 92281 1
  1. Mathematical statistics. I. Conrad, Simon.
QA276.16.A77 1989
519.5—dc20   89-31594
                                                      CIP

British Library Cataloguing in Publication Data:
Assignments in applied statistics.
  1. Statistical mathematics
  I. Conrad, Simon
  519.5

  ISBN 0 471 92281 1

Typeset in Northern Ireland by The Universities Press (Belfast) Ltd.
Printed and bound in Great Britain

Contents

Contributors

S. Conrad (editor)
School of Management, University of Manchester Institute of Science and Technology, PO Box 88, Manchester M60 1QD

After degrees in mathematical statistics and in operational research, and two years with the Government Statistical Service, Simon Conrad joined what is now the School of Management at UMIST. His recent research and publications have concerned the use and need for quantitative analysis within organizations, and the professional development of statisticians and operational researchers. He is a partner of a consultancy specializing in current awareness programmes in management science within Europe, and Editor of *Journal Contents in Quantitative Methods.*

R. Caulcutt
Management Centre, University of Bradford, Emm Lane, Bradford, West Yorkshire BD9 4JL

Roland Caulcutt is the BP Chemicals Lecturer in Statistical Process Control at the University of Bradford. Previously he worked as a statistician with ICI and later served many other companies within the chemical industry as an independent consultant. He has taught applied statistics, electronic engineering, quality management and educational psychology in a variety of institutions. He is the author of several texts in statistics for scientists and technologists.

G. Dunn
Biometrics Unit, Institute of Psychiatry, De Crespigny Park, Denmark Hill, London SE5 8AF

Graham Dunn originally trained as a biochemist and, after four years' postdoctoral research at the University of Oxford, abandoned laboratory work in favour of statistics. Having obtained an MSc in Applied Statistics at Oxford in 1979, he obtained a post as a statistician in the General Practice Research Unit at the Institute of Psychiatry. Since 1983 he has been a member of the academic staff of the Institute of Psychiatry's Biometrics Unit. He is currently a senior lecturer in statistics in the Unit.

B. S. Everitt
Biometrics Unit, Institute of Psychiatry, De Crespigny Park, Denmark Hill, London SE5 8AF

Brian Everitt is Head of the Biometrics Unit at the Institute of Psychiatry, a research school of the University of London, and director of Sigma X, a statistical and data analysis consultancy based in London. His main statistical interests are in multivariate analysis, particularly cluster analysis and mixture distributions. He is the author of nine books on statistics ranging from *Cluster Analysis* to *Statistical Methods in Medical Investigations.*

D. J. Hand
Department of Statistics, The Open University, Walton Hall, Milton Keynes MK7 6AA

After eleven years at the London University Institute of Psychiatry, advising researchers from many different disciplines on how to collect and analyse their data, as well as teaching on the graduate programme, David Hand took up the Chair of Statistics at the Open University. He is also Chairman of Sigma X, the statistical and data analysis consultancy. His research interests include multivariate statistics and statistical expert systems, and his publications include eight books in these areas.

Z. W. Kmietowicz
School of Business & Economic Studies, University of Leeds, Leeds LS2 9JT

After industrial experience in operational research, Zbigniew Kmietowicz joined the University of Leeds in 1964, as lecturer in economic statistics. His research interests span two main fields: decision making under uncertainty and the economic statistics of developing countries. As well as having undertaken several major consultancies in Africa for international development agencies, Mr Kmietowicz has held visiting university appointments in Kenya and Uganda.

A. D. Pearman
School of Business & Economic Studies, University of Leeds, Leeds LS2 9JT

Alan Pearman is senior lecturer in the School of Business and Economic Studies at the University of Leeds. His research interests concern the use of quantitative methods to guide decision making, particularly in the public sector. He is an associate lecturer in the University's Institute for Transport Studies and has held visiting appointments at the universities of Bradford, Nanjing and North Carolina.

J. Q. Smith
Department of Statistics, University of Warwick, Coventry CV4 7AL

James Q. Smith obtained a first class mathematics degree at the University of Nottingham before completing a PhD in Mathematical Statistics at the University of Warwick in 1978. He spent five years as a lecturer at University College, London, before moving to the University of Warwick. He has published widely in Bayesian decision theory, forecasting, and graphical representations of conditional independence. He is the author of *Bayesian Decision Analysis* and an active member of the University of Warwick Statistical Consultancy Unit.

J. R. Sparkes
Management Centre, University of Bradford, Emm Lane, Bradford, West Yorkshire BD9 4JL

John R. Sparkes is Assistant Director of the University of Bradford Management Centre and Senior Lecturer in Managerial Economics. He has published several books and written numerous articles in prominent economics and management journals, many on forecasting. He is a founder member of the Editorial Advisory Board of the international journal of research in management economics, *Managerial and Decision Economics*.

S. Stray
School of Industrial & Business Studies, University of Warwick, Coventry CV4 7AL

Since graduating from the University of York, Stephanie Stray worked at both Bradford Management Centre and Aston Business School. In 1985 she joined the staff of the Operational Research and Systems Group at Warwick Business School. In recent years she has obtained a PhD from the University of Essex, has been engaged in research upon the analysis of survey data, opinion poll data and upon British Parliamentary by-elections, and is a consultant to OECD.

G. J. G. Upton
Department of Mathematics, University of Essex, Wivenhoe Park, Colchester CO4 3SQ

After five years as a lecturer in the Department of Statistics at the University of Newcastle-upon-Tyne, Graham Upton moved to the University of Essex in 1973 where he is currently senior lecturer. He is the author of more than fifty papers dealing with various aspects of applied statistics, and of three books, the most recent of these being the two volumes of *Spatial Data Analysis by Example* (with co-author B. Fingleton).

S. F. Witt

Department of Management Science & Statistics, University College of Swansea, Singleton Park, Swansea SA2 8PP

Stephen F. Witt is Lewis Professor of Tourism Studies in the Department of Management Science and Statistics at the University College of Swansea. He is co-author of *Portfolio Theory and Investment Management, Practical Business Forecasting, Practical Financial Management* and the *Tourism Marketing and Management Handbook*. He has published widely in academic journals, with particular emphasis on demand forecasting, and is a member of the Editorial Board of *Tourism Management*.

Preface

Assignments in Applied Statistics is written for statistics courses in mathematics and statistics departments, where applications-orientated material is considered relevant and useful, and for statistics courses, more naturally applications-orientated, in management and business departments, and social sciences, natural sciences, engineering and other departments. Many of the topics covered will also be found in management and MBA programmes in business schools, and in some courses in operational research.

It is assumed that students using *Assignments in Applied Statistics* will have completed an earlier introductory course in statistics, together with some later introductory lectures on the subject covered by the assignment in question. There has necessarily been an element of subjectivity in the choice of four subjects under each of five applied statistics headings, but generally they are among subjects that would be covered by introductory lectures in *Decision Analysis, Forecasting, Multivariate Analysis, The Design and Analysis of Surveys,* and *The Design and Analysis of Experiments.*

The lecturer would use one of the assignments to test his (or her) students' understanding of the subject. The students, apart of course from being told to do the assignment, would be looking forward to some practical experience in data handling and analysis and, if they have not understood the lecturer, to some references to some further reading!

I became aware some years ago, from my own teaching and from informal discussions with colleagues, that there was little readily available material that would provide practice in data handling and analysis, and that could also be used as course assessment in statistics. Also that lecturers often faced severe time constraints, and were often obliged to move on from a particular subject without feeling they had done it justice. The assignments are designed to at least partially alleviate this frustration, in that they encourage, and indeed sometimes demand, the student to read around the subject.

These are assignments, not case studies. Although experience of problem formulation is no doubt worthwhile, there are other considerations. Indeed some would argue that problem formulation should be learned 'on the job', that the lecture room should be used for teaching concepts and techniques; others that the 'process' of statistics is not easily teachable or well-defined, or does not even exist.

With these assignments there is little problem formulation. The aim is to give students practice in data handling and analysis, and to introduce them to the different aspects of the subject which may not have been covered by the lecturer.

Any lecturer introducing a particular subject will most probably cover roughly the same ground, but there can be no guarantee. To make each assignment useful and relevant to as many readers as possible, references within each assignment range from those which are introductory and provide a general overview of the subject, to references which are specific to particular points made or questions asked. No assignment, therefore, should in the main be beyond the capability of the average student; simply, the less the coverage by the lecturer, the more reading and the more work required from the student.

The assignments therefore provide an incentive for students to visit libraries, search for material, and finish with a more rounded appreciation for their subject. Students in quantitative subjects, in statistics or indeed operational research, have too often, I believe, been able to rely on lecture notes and worked examples provided by lecturers, and have not found it necessary to frequent libraries or read articles in academic journals, the most useful of which will quite possibly be applications-orientated. It is this, I believe, that has on occasions made quantitative students less able to place their knowledge and understanding within a context than some other students, and poorer for it.

The assignments have all been specifically written for this text, in order to meet the above objectives. Many are based on the authors' earlier investigations and practical experiences, and have been adapted to allow the reader to develop the subject for him- or herself. Each is set within a different applied context, but in a general way and certainly not to the extent that the context masks the statistical principles.

The assignments have been grouped into five sections, each consisting of an Introduction followed by four Assignments. Each introduction provides a brief overview of the section, and introduces the assignments. Each assignment then starts with an outline and a list of keywords, in order for the reader to more easily assess the relevance and appropriateness of any particular assignment.

Generally the assignments have been written to a common format (not a task to be underestimated with eleven different contributors), subject only to the particular requirements of individual assignments. Depending on what is appropriate, the questions are either interspersed within the assignment or appear together towards the end of the assignment, with the references. There then follow a number of supplementary questions, followed by any further references necessary for these questions. Depending on the assignment, and the requirements of the reader, some of the questions may be omitted (or questions added); certainly the supplementary questions are optional.

A Teachers' Guide is available from the Editor, and contains outline answers to the questions and supplementary questions.

I would like to thank the many academics throughout the UK who, at different times, and sometimes without even realizing it, helped to establish the parameters for this text. Also the publishers for their continued support, given the time that has elapsed since conception. Finally the contributors themselves, without whom of course the text would not have been possible. And on behalf of any one contributor, I extend thanks to the others.

Simon Conrad

Manchester
October 1988

DECISION ANALYSIS

- Introduction
- Utility and Probability Assessment in Settling an Insurance Claim
- Multicriteria Analysis and Public Sector Decision Making
- A Decision Tree for the Selection of Tenderers
- Sample Information in Marketing a New Product

Assignments in Applied Statistics
Edited by S. Conrad
© 1989 John Wiley & Sons Ltd

Introduction

Simon Conrad

*School of Management, University of Manchester Institute of
Science and Technology*

The phrases decision making, decision theory and decision analysis tend at times to be used almost interchangeably by different authors, yet we can attempt to give a different meaning to each. Decision making can be interpreted either as how we make a decision or as how we should make a decision. The former is behavioural, how an individual or group of individuals actually arrives at a decision; the latter is normative, how individuals or groups should act in order to act rationally. Decision theory can be described as a normative approach to decision making which is essentially theoretical. Decision analysis is also normative, but is concerned with applications and might be described as applied decision theory.

It is decision analysis that concerns us here: how we might use different concepts and techniques to improve our decision making. It is true that much can be construed as decision making, from an individual deciding whether or not to buy this book to a company deciding whether or not to enter a new market. However, it is only when the consequences which would arise from a particular decision are perceived to be sufficiently important that we stop to consider whether we should formalize the problem before us. Most probably there is at least some deliberation when deciding whether or not to buy a book, but it will almost certainly be momentary, and will not be formalized. The consequence is that we are not certain we have considered all the actions open to us, all the possible consequences; in other words we are not really sure we have acted rationally.

Part of the problem is that we are not sure that the time and energy expended in formalizing the problem we face is going to be worthwhile, that any gain is going to exceed the cost of formalization. There is in fact an analogy in decision analysis itself. Situations arise where we have to decide between making a decision immediately, on the basis of the information that is currently available, or alternatively waiting until more information is available, perhaps by performing an information-gathering experiment, and making a decision on the basis of this further information. As above, is the gain from having this additional information to aid our decision making going to exceed the cost of the delay or the experiment?

It is this formalization of the problem in decision analysis that can present us with some of the greatest difficulties. Problem formulation can often be more difficult than problem analysis. It is a truism that a problem with a simplistic formulation stands a greater chance of solution, but also a greater chance of being criticized for its simplicity, an abstraction of the real-life situation it is supposed to represent. Generally, we aim for a formulation that is as realistic (and therefore as complicated) as we can handle.

The simplest formulation requires the structuring of the problem so that there are a number of actions to choose from, a number of (unknown) states of nature, utilities associated with the consequences of choosing a particular action given a particular state of nature, and a probability assessment over the states of nature. In this case, with a finite number of discrete states of nature, the problem will be called discrete.

Many decision analysis problems are, however, naturally continuous (for example, the state of nature being the unknown percentage defective for a particular manufacturing process or the unknown future sales level of a particular product). Continuous problems are most often solved by reformulating them as discrete problems. This involves dividing the range of a (continuous) state of nature into a finite number of discrete intervals. The assignments in this section essentially deal with discrete problems.

In decision analysis, some or all of the following issues typically arise, that is problems have a combination of some or all of the following characteristics. Each is considered at length in the decision analysis literature, and in fact both multidimensional and sequential problems are or can be sufficiently complex for them to be considered topics in their own right.

Decisions may be taken by reference to either non-probabilistic (for example, maximin utility) or probabilistic decision criteria, in particular the maximization of expected utility (MEU) criterion. Choosing an action by reference to a particular non-probabilistic decision criterion is unfortunately not particularly satisfactory, since in general different criteria lead to different actions, and the problem is simply translated from choosing an action to choosing a decision criterion.

Using the MEU criterion does, however, require a probability assessment. If the problem is discrete, then a probability distribution will usually be required, and if continuous, then a probability density function. (The actual assessment is not necessarily in one of these forms, and the distribution or density function may have to be derived.) This assessment may be objective, that is based on historical data, but in some instances historical data will not exist. In this case a subjective assessment by the non-specialist involved in and familiar with the problem will be necessary, and in some instances this will actually be preferred to any objective assessment. Eliciting a (subjective) probability assessment from a non-specialist is obviously an important task and a number of well-documented procedures exist. In some instances a complete probability assessment may not be possible or available: we may, for example, have to manage with only a ranked order of the probabilities.

In many practical situations, utility is a multidimensional function, for example the utility a firm derives from a particular marketing strategy will be a function of both profit and market share. Frequently, however, utility is assumed to be a function of just one variable (most commonly, profit). In either case, a utility function would normally have to be assessed.

However, if in the case of a single variable the function is linear (at least over the profit range for the problem), then the maximization of expected profit (MEP) and the MEU criteria are equivalent. It is for this reason that the MEP criterion is commonly referred to, in other words the above assumptions are implicit.

Utility functions can also be assessed directly, for in some situations we may not find it easy to define exactly what variables should appear in the utility function. In either case, whether a utility function is constructed from a number of formally specified variables or is assessed directly, the usual procedure is one of a number of cross-questioning techniques, between decision analyst and client.

The first assignment, on deciding whether or not to settle an insurance claim out of court, requires that the insurance company specifies an appropriate utility function and a probability distribution of the eventual further costs to the company of allowing the legal argument to continue. The information provided is in the form of a question and answer session between the decision analyst and the company spokesman, and is typical of the way in which utility and probability assessments are made.

Many of the reported applications of decision analysis are concerned with multidimensional problems. Phrases such as multiattribute utility, multiple criteria decision making and decisions with multiple objectives all appear in the literature. Multidimensional problems in decision analysis, and particularly where the objectives are conflicting, have much in common with cost–benefit analysis (in economics) and with goal programming (in operational research).

The second assignment deals with an application of multicriteria decision analysis to public sector decision making. Here the concern is with choosing one of a number of different economic development strategies, each one affecting a particular city in a number of different ways: for example, through effects on employment levels, average per capita income and pollution. These different factors are in effect the criteria by which the strategies will be ranked and a preferred strategy chosen. The assignment examines the different procedures that exist for choosing a preferred strategy.

Problems involving some or all of the above issues can be formulated, analysed and optimal actions determined. When an action is chosen on the basis of an initial probability assessment, whether it be subjective or objective, and without any further information either arising naturally or being collected, then we have a prior analysis of the problem, and the distribution or density function is called a prior distribution.

In many situations, however, a decision can be delayed until more information becomes available. This information can, at least in theory, be used to revise the prior distribution through Bayes' theorem, the revised distribution being called a posterior distribution. If the MEU criterion is based on this posterior distribution, then we have what is called posterior analysis. Prior analysis can therefore be interpreted as prior to (or before) any further information and posterior analysis as posterior to (or after) any further information.

Pre-posterior analysis allows the decision maker to determine whether or not a delay or a particular information-gathering experiment is worth performing, and also to choose from different experiments. It is worth noting here that 'different experiments' can either mean a number of distinctly different experiments, for example choosing one from a number of different diagnostic medical tests, or alternatively different sizes of what is essentially the same experiment, for example choosing the number of items to be inspected in a sampling inspection scheme.

Certainly the concept of optimal experimentation is very useful: however, it is important to appreciate that Bayes' theorem for pre-posterior analysis (and, to a lesser extent, for posterior analysis) is not easily applied for continuous problems. There are, admittedly, certain well-documented situations where an analytical solution is possible, but these are hedged with restrictive assumptions. The more useful procedure, already referred to, is to approximate a continuous problem by a discrete one.

Obviously problems can arise where the nature of a second and further experiment is determined by the outcome of the first experiment. Such multistage or sequential problems can be solved either through the use of decision trees or, alternatively, and depending on the nature and complexity of the problem, by dynamic programming or decision networks.

In the third assignment, a manager has to decide which of a number of possible contractors he will ask to tender for two maintenance contracts. The optimal tendering policy is to be determined from a decision tree. Certain necessary assumptions can be represented by an influence diagram, to aid understanding. The possibility of an analytical solution and the generalization of the problem, involving the pruning of what becomes a 'bushy tree', are also considered.

In the fourth assignment, a company has to decide whether or not to market a new product. Information from a market research bureau is available at a price, and the company has to decide whether or not to purchase this information. This is a 'classic' pre-posterior analysis problem, and here the solution is through a decision tree. A related problem involving a continuous state of nature is also investigated, together with one where the company has 'incomplete information', being able only to rank probabilities rather than specify them exactly.

Collections of articles on decision analysis appear in Kaufman and Thomas (1977) and Howard and Matheson (1985). Examples from

Kaufman and Thomas are Hespos and Strassmann's *Stochastic Decision Trees for the Analysis of Investment Decisions*, Bierman and Hausman's *The Credit Granting Decision* and Abernathy and Rosenbloom's *Parallel Strategies in Development Projects*. Pearman (1987) reports on the findings of a survey of leading exponents of decision analysis in the United Kingdom and the United States.

A number of texts on decision analysis have been published recently, including Buchanan (1982), Bunn (1984) and Smith (1988). Moore and Thomas (1976) is a useful introductory text.

An overview of decision analysis is provided by Keeney (1982). Winkler (1982) and, more recently, Moskowitz and Bunn (1987) discuss research opportunities and current trends in decision analysis. Henrion (1985) reports on the availability of software for decision analysis.

REFERENCES

Buchanan, J. T. (1982). *Discrete and Dynamic Decision Analysis*, Wiley, Chichester.

Bunn, D. W. (1984). *Applied Decision Analysis*, McGraw-Hill, New York.

Henrion, M. (1985). Software for decision analysis, *OR/MS Today*, **12**, 24–9.

Howard, R. A., and Matheson, J. E. (eds.) (1985). *Readings on the Principles and Applications of Decision Analysis*, Strategic Decisions Group, Menlo Park, California.

Kaufman, G. M., and Thomas, H. (eds.) (1977). *Modern Decision Analysis*, Penguin, Harmondsworth, Middlesex.

Keeney, R. L. (1982). Decision analysis: an overview, *Operations Research*, **30**, 803–38.

Moore, P. G., and Thomas, H. (1976). *The Anatomy of Decisions*, Penguin, Harmondsworth, Middlesex.

Moskowitz, H., and Bunn, D. (1987). Decision and risk analysis, *European Journal of Operational Research*, **28**, 247–60.

Pearman, A. D. (1987). The application of decision analysis: a US/UK comparison, *Journal of the Operational Research Society*, **38**, 775–83.

Smith, J. Q. (1988). *Decision Analysis: A Bayesian Approach*, Chapman and Hall, London.

Winkler, R. L. (1982). Research directions in decision making under uncertainty, *Decision Sciences*, **13**, 517–33 (and discussion, 534–53).

Assignments in Applied Statistics
Edited by S. Conrad

Utility and Probability Assessment in Settling an Insurance Claim

J. Q. Smith

Department of Statistics, University of Warwick

OUTLINE

A company has to decide whether or not to settle an insurance claim out of court. A company representative is interviewed about preferences and beliefs. You must use the ensuing information to elicit the utility function and probability distribution and hence to advise as to the best course of action. You are finally asked to describe how you would conduct another such interview if your client's utility function were multiattribute.

KEYWORDS

preferences, beliefs, utility function, probability distribution, elicitation, incoherence, optimal policy, measurement error, multiattribute utility

When an insurance company receives claims of a substantial sum, the company will often contest the amount claimed through the courts. This action, however, is very costly to the company and so, instead of fighting the court case to the finish, and depending on how the case develops, it may offer a settlement to the claimant out of court.

In practice the company often makes interim payments to a claimant during the court case. It may make many offers of settlement through the duration of a case and, of course, the claimant may well choose to decline any offer of settlement made. However, for the purpose of this assignment, to keep the problem more tractable, we shall assume that it has been made clear to the insurance company by the claimant's lawyers that they will not accept an immediate settlement of less than £210,000. On the other hand, if such an offer is made their client will accept this payment out of court. We shall also assume that it will not be viable for the client to offer a later settlement when the court case has proceeded further, or that any interim payments will be made.

Insurance companies usually have legal experts who, on the evidence they are presented with from the court action, make an informed decision on whether or not to offer such a settlement. In the case described here the company has employed a decision analyst to help and rationalize the way the expert will come to such a decision. To give his advice the analyst needed to interview the client to elicit both his utility function and his probability distribution of the eventual further costs to the company of allowing the legal argument to continue. It was very fortunate that this expert was very numerate and interested in formalizing the mechanism by which he came to his decisions.

Two précised and adapted transcripts of parts of the interview between the expert (E.) and decision analyst (D.A.) are given below. Again, for the sake of clarity, much of the discourse and technical points have been omitted. For the purpose of this assignment, questions about his utility function and his probability assessments of uncertain quantities have been recorded separately.

THE UTILITY TRANSCRIPT

D.A. To summarize the information you have given me so far: you tell me that it is not possible for you to gain anything from the action, at best you will be awarded your legal costs. On the other hand, taking all legal costs into account you tell me that the probability that the total cost to the company of the case will be more than £1,000,000 is negligible. You also tell me that your current legal expenditure is £50,000.

E.   That is correct.

D.A.   Now suppose you knew that there was a 50–50 chance that the court case would cost you either $x - h$ or $x + h$ (in £s) where both these amounts are between 0 and one million £s. For various values of x and h I want you to estimate the amount you believe the company should be prepared to pay to settle the claim.

E.   It is difficult to evaluate how I should act then; these types of options are not offered in practice. I think it would be fair to say that we would be prepared to pay somewhat more than the average of these two amounts (x). I would also worry if company policy depended on the average value (x) of these possibilities just because of the ambiguous nature of the term 'cost' in this context. Do you mean cost over and above the amount for which we have budgeted, net cost or what? Certainly the company will assess its performance against targets as yet incompletely specified and its preferences will be governed by these.

D.A.   You have put your finger on a tricky problem here, but to answer your question, I mean *net* cost. Suppose you were to assess that there was a 50–50 chance that the court case would cost you nothing or a million £s. How much would you be prepared to settle for in such a case?

E.   Well, this option would be extremely unlikely you know—we haven't paid out a claim of that quantity in the last ten years.

D.A.   I understand, but can you answer the hypothetical question?

E.   Well, I suppose somewhere between £550,000 and £650,000—say approximately £600,000.

D.A.   So you believe that the maximum amount the company should envisage losing on a settlement is £600,000 plus the costs of the case so far, that is £650,000.

E.   Yes, approximately, though this is a guess.

D.A. I appreciate that; you can modify these judgements as we continue if you like.

The interview proceeded with the decision analyst asking the expert about his maximum acceptable settlement for several such 50–50 gambles. The results of this part of the interview for these 50–50 gambles, some modified by the expert after reflection, are given in Table 1. (The client was not confronted with any apparent incoherence between his stated preferences and an expected utility maximizing strategy at this stage.)

An introduction to the elicitation of subjective utilities can be found in, for example, Moore and Thomas (1976, Chs. 9 and 10), Buchanan (1982, Ch. 3), Bunn (1984, Ch. 3), Lindley (1985, Chs. 5 and 9) and Smith (1988, Ch. 3). More detailed information is given by Swalm (1966), Hull, Moore and Thomas (1973), Tversky and Kahneman (1974), Johnson and Huber (1977) and Tversky (1977).

Table 1
Maximum acceptable settlement cost as stated by the client, equivalent to various gambles

QUESTION NUMBER	50–50 GAMBLE ON NET COST OF $x-h$ AND $x+h$ (in £m)		MAXIMUM SETTLEMENT (+CURRENT COSTS) ACCEPTABLE (£m)
	$x - h$	$x + h$	
1	0	1	0.65
2	0	0.65	0.35
3	0.65	1	0.85
4	0.35	0.85	0.6
5	0	0.35	0.1
6	0.35	0.65	0.45
7	0.35	1	0.725
8	0	0.85	0.5
9	0.65	0.85	0.7
10	0.85	1	0.95
11	0.1	0.95	0.55
12	0.45	0.725	0.6
13	0.1	0.45	0.3
14	0.1	0.725	0.425
15	0.45	0.95	0.7
16	0.725	0.95	0.85

THE PROBABILITY TRANSCRIPT

D.A. What I now intend to do is to quantify what you believe will happen. To do this I need to elicit the probabilities that you are implicitly or explicitly assigning to various events concerning the likely outcome of the case you are fighting. Do you have any documented evidence of what happened in previous cases similar to this one?

E. Yes, we have a great deal of such evidence to hand. Of course, every case is different from every other but in very loose terms I have found 37 cases documented which could be judged to be similar to this one where we fought the case to the finish and lost. I have calculated the net cost x to the company (claim payment plus both the expenses of the company and those of the claimant) of these similar cases. To make the amounts more relevant the figures have been adjusted upwards to take account of inflation over the last three years. (This distribution is given in Table 2.)

D.A. Good, this information will be very valuable to us. What I need to do now is to find out what you believe will happen in the court case.

E. Well, clearly we will either win or lose the case. If we were to win the case we would almost certainly be awarded costs so our total

Table 2
Costs (adjusted for inflation) for similar successful claims made in the last three years

COST (£m)	FREQUENCY
0–0.1	1
0.1–0.2	0
0.2–0.3	5
0.3–0.4	10
0.4–0.5	14
0.5–0.6	6
0.6–0.7	1
0.7–1.0	0

expenditure would be nothing. On the other hand, if the company lost the case, unless very unusual circumstances arise, and I don't think they will here, the company will be required to pay the claimant's legal costs as well as its own.

D.A.   Both from the information you have given me (Table 2) and what you have said just now, it is clear that we should break down the distribution of the cost to the company of the action, separating the costs when you win from the costs when you lose.

E.   That seems sensible—costs are obviously zero if we win, provided we are awarded costs of course.

D.A.   Yes, I understand that. So what I need to ascertain from you is the probability that you will win this particular case. Would you judge it to be less than a 50–50 chance?

E.   Well, let met see. Obviously I would *like* it to be more than a 50–50 chance, but I would be sticking my neck out if I were to make this judgement. The company obtains a transcript of this interview and it will look bad for me if I am seen to be over-optimistic. So I will say that we have a 40 per cent chance of success.

D.A.   So if I were to offer you the hypothetical option of your winning or losing the case depending on whether you pulled a red or black ball out of an urn containing four red balls and six black balls you would be indifferent between this hypothetical option and the real one.

E.   Well, I find this comparison very artificial. I find I am much happier with the type of hypothetical option you offer because at least there I *know* the relevant probabilities. The real problem is more difficult because I do not *know* the relevant probabilities. But to answer your question I suppose I find each option equally preferable. Certainly on our past record we have won less than half such cases.

D.A.   Well, this figure of 40 per cent can be adjusted later. All we need to ascertain now is your probability of the net cost to the company were you to lose the case.

E.   Well you see the information in front of you (Table 2). I believe this is a fairly representative sample of costs arising from cases such as this one.

D.A. So you would be happy to use this distribution for your distribution of total costs given that you lost the case.

E.   Yes.

An introduction to the elicitation of subjective probabilities can be found in, for example, Moore and Thomas (1976, Chs. 7 and 8), Buchanan (1982, Ch. 2), Lindley (1985, Ch. 9) and Smith (1988, Ch. 4). More detailed information is given by Hampton, Moore and Thomas (1973) and Lindley, Tversky and Brown (1979). Some documentation on beliefs of the practitioner in real-life problems is Grayson (1960) and Spetzler (1968).

After the interview the decision analyst went away to calculate an appropriate utility function and subjective probability distribution for the outcome of the case on the basis of the information given above. These, together with the corresponding optimal policy, were discussed with the expert. The expert was particularly interested in the effects of changing various model assumptions made by the analyst.

Q1 Comment on the ways the two interviews are conducted. Can you make any suggestion about how they might be improved?

Q2 Convert the expert's statements about his company's preferences (which are summarized in Table 1) into statements about his company's utility function. Are you able to explain any 'measurement errors' you find associated with the elicitation procedure?

Q3 Find a utility function for the net cost C to the company (in £m) of the form

$$U(C) = 1 - \left(\frac{e^{sC} - 1}{e^s - 1} \right)$$

where ($s > 0$ is a parameter) which is a good approximation to the expert's utility function.

Q4 It is suggested that the distribution of C, the net cost of the case if the company lose, is Normal. Estimate the distribution of C under this assumption together with an assumption made by the expert during the interview. Can you think of any reason why the expert's assumption may be suspect?

Q5   On the basis of the modelling assumptions in Questions 3 and 4 above, instruct the expert in whether or not he should settle immediately. Does this decision also minimize expected costs for the company?

Q6   Clearly the parameters you have estimated from this interview will be subject to certain 'measurement error'. Investigate how robust your result is to misestimation of the parameters in your model, paying special regard to those types of misestimation you might expect from the proceedings of the interview.

REFERENCES

Buchanan, J. T. (1982). *Discrete and Dynamic Decision Analysis*, Wiley, Chichester.

Bunn, D. W. (1984). *Applied Decision Analysis*, McGraw-Hill, New York.

Grayson, C. J. (1960). *Decisions under Uncertainty: Drilling Decisions by Oil and Gas Operators*, Division of Research, Harvard Business School, Boston, Massachusetts.

Hampton, J. M., Moore, P. G., and Thomas, H. (1973). Subjective probability and its measurement, *Journal of the Royal Statistical Society (Series A)*, **136**, 21–42.

Hull, J., Moore, P. G., and Thomas, H. (1973). Utility and its measurement, *Journal of the Royal Statistical Society (Series A)*, **136**, 226–47.

Johnson, E. M., and Huber, G. P. (1977). The technology of utility assessment, *IEEE Transactions on Systems, Man, and Cybernetics*, **SMC-7**, 311–25.

Lindley, D. V. (1985). *Making Decisions*, Wiley, Chichester.

Lindley, D. V., Tversky, A., and Brown, R. V. (1979). On the reconciliation of probability assessments (with discussion), *Journal of the Royal Statistical Society (Series A)*, **142**, 146–80.

Moore, P. G., and Thomas, H. (1976). *The Anatomy of Decisions*, Penguin, Harmondsworth, Middlesex.

Smith, J. Q. (1988). *Decision Analysis: A Bayesian Approach*, Chapman and Hall, London.

Spetzler, C. S. (1968). The development of a corporate risk policy for capital investment decisions, *IEEE Transactions on Systems Science and Cybernetics*, **SSC-4**, 279–300.

Swalm, R. O. (1966). Utility theory—insights into risk taking, *Harvard Business Review*, **44**(6), 123–36.

Tversky, A. (1977). On the elicitation of preferences: descriptive and prescriptive considerations. In D. E. Bell, R. L. Keeney and H. Raiffa (eds.), *Conflicting Objectives in Decisions*, Wiley-Interscience, Chichester.

Tversky, A., and Kahneman, D. (1974). Judgement under uncertainty: heuristics and biases, *Science*, **185**, 1124–31.

SUPPLEMENTARY QUESTIONS

1. If C has a Normal distribution, find the greatest value of settlement costs p^* the company should be prepared to pay as a function of:

   (a) the mean m and variance V of costs C if the case is lost;
   (b) the expert's assessment of the probability $q > 0$ that he loses the case;
   (c) the parameter of his utility $s > 0$.

   Prove that the total settlement costs p will always be the more acceptable than continuing to fight the case if

   $$p \leqslant m - s^{-1} \log\left(\frac{1}{q}\right) + \tfrac{1}{2}sV$$

2. Repeat the analysis above when, instead of assuming that C has a Normal distribution, assume C has a Gamma distribution with the same mean and variance. Calculate p^* as a function of your other parameters, as you did above, and show that for some fixed values of s and V/m settlement is preferred for all values of m. Interpret this result.

3. There are many times when there is little or no hard data as there was in this example on which the expert can base probability judgements. If, in this example, the distribution of C, the cost of an unsuccessfully fought case, could not be directly related to data, discuss at least two different ways in which this distribution could be elicited from a client. Outline some of the drawbacks of either method.

   Draw up a transcript between an imaginary expert and yourself which would indicate how you would hope such an interview to progress if you used one of the methods you outline above. Discuss how you might resolve any conflict in information that might occur in this elicitation procedure.

4. An insurance company has not only to consider its immediate financial position when making a decision but also the effect that bad publicity in a particular court case will have on the sales of their policies and hence their longer term financial position. The client wants to include this factor in the analysis.

   You decide to try to elicit a biattribute utility function from your client, the two attributes being:

   (a) immediate financial efficacy and
   (b) the effect of publicity on the future financial position.

   Assuming that these two attributes are utility independent of each other, describe how you might elicit a utility function from your client. Can you see any reason why the independence assumption made above might be broken? Describe how you might attempt to improve utility elicitation in the light of this.

Two books that discuss multiattribute decision making are Keeney and Raiffa (1976) and Bell, Keeney and Raiffa (1987).

5. Further to Supplementary Question 4, your client also states that he believes that the company should not be seen to be 'too soft'; if they do, they expect to have claimants fighting claims through the courts in the confident expectation that by doing this they will 'bully' the company into making large settlements out of court. How could this new judgemental criterion be incorporated into the company's utility function?

FURTHER REFERENCES

Bell, D. E., Keeney, R. L., and Raiffa, H. (eds.) (1977). *Conflicting Objectives in Decisions*, Wiley-Interscience, Chichester.

Keeney, R. L., and Raiffa, H. (1976). *Decisions with Multiple Objectives: Preferences and Value Tradeoffs*, Wiley, New York.

Assignments in Applied Statistics
Edited by S. Conrad
© 1989 John Wiley & Sons Ltd

Multicriteria Analysis and Public Sector Decision Making

A. D. Pearman

School of Business & Economic Studies, University of Leeds

OUTLINE

This assignment is concerned with decision making in a complex environment, characterized by the wish both to reconcile the alternative viewpoints of the various parties to the decision and to take account of the many different ways in which public policy can affect the life of a community. To be faced with decisions of this type is common in the public sector. The methodological basis of the assignment is drawn from the growing range of formal, quantitative multicriteria decision aids. Particular emphasis is placed on the use of formal tools to increase decision makers' insights about the nature of the choice to be made. In the supplementary questions, opportunities are provided to expand the basic methodology using multiple regression analysis and linear programming.

KEYWORDS

Multicriteria methods, uncertainty, modelling, multiple attribute decision making, value function, attributes, problem structuring, value tree, SMART (simple multiattribute rating technique), SAW (simple additive weighting) value function, ratio technique, weights, dominance, sensitivity testing, multiple regression, linear programming, cost–benefit analysis

This assignment is concerned with a problem of decision making in the public sector. It is not uncommon for the application of quantitative tools to 'management' problems to be taken to be the exclusive concern of the private sector. That is far from being the case. There are many important public sector decisions that are particularly well suited to benefit from the application of quantitative techniques. The present assignment, although it relates to a hypothetical example, is typical of a range of similar policy choice problems which have been aided by the application of multicriteria methods. (See, for example, Keeney, 1982, pp. 824–5); Voogd, 1983, Chs. 14–16; Winterfeldt and Edwards, 1986, Ch. 12.)

The Economic Development Authority (EDA) of the city with which the assignment is concerned is under pressure to implement an economic development strategy. Officers of the authority have spent some time researching possible strategies and have arrived at a shortlist of five. As with many socioeconomic policy options, the available strategies will potentially affect the city in several different ways, for example through effects on employment levels, average per capita income, pollution, etc.

The task is to prepare a report to present to a committee of elected political representatives drawn from the city council which will:

(a) set out the principal characteristics which differentiate the five strategies;
(b) make an EDA recommendation as to which one is the most attractive to implement;
(c) provide the elected representatives with a framework for discussing the officers' recommendation in the light of their own views about the relative importance of the different impacts of the strategies.

Responsibility for the final choice of strategy rests with the political representatives.

The information in terms of which the five strategies will formally be assessed is set out in Table 1. In a real-world decision making exercise, a

Table 1
Impacts of the five economic development strategies

STRATEGY	IMPACT							
	1	2	3	4	5	6	7	8
A	23	0.8	3.0	14	45	25	4	60
B	18	0.4	0.9	2	35	10	1	10
C	21	0.5	3.0	2	20	5	1	10
D	50	2.4	6.2	42	20	65	8	100
E	33	1.6	2.4	27	25	5	1	10

considerable amount of analysis must be undertaken before such data can be specified. These steps will be discussed briefly on pages 23–25. The data are introduced now in order to give a clearer picture of the strategies to be compared and to emphasize that what is required is a careful assessment of the relative achievements of the five strategies in terms of dimensions which embrace a wide range of different types of impact, measured in a correspondingly wide range of units. The different dimensions provide the criteria through which the strategies will be ranked and a preferred strategy chosen. Methods which help decision makers to take information like that in Table 1 and compare alternatives in terms of all their different impacts simultaneously are termed *multicriteria decision methods*. Note that the assessed impacts of the strategies reflect not only the nature of the strategies themselves but also the EDA's estimate of how effective they might eventually be.

Impacts

1. Increase in local income (£m p.a.)
2. Extra tax and other revenues accruing to the city (£m p.a.)
3. Necessary extra investment in public infrastructure (£m annuitized)
4. Demand for unskilled labour (hundreds of workers)
5. Contribution to broadening the city's economic base (index 0–100)
6. Percentage of vacant industrial land occupied
7. Extra heavy vehicle movements (100,000s per year)
8. Increase in pollution (index 0–100)

Strategies

A. Small firms: concentrate aid on providing facilities and advice for small local businesses
B. High technology: promote a 'science park' in conjunction with local higher education institutions
C. Community image: no direct focus in terms of attracting particular industries; promote the public image of the area and provide general advice and support to firms expressing interest in relocation
D. Manufacturing industry: designate sites for large traditional manufacturers, provide good transport access and try to attract relocating firms or foreign investment
E. Finance and administration: promote the area as a regional financial and administrative centre, with good passenger travel facilities, telecommunications, etc., and a well-trained workforce.

BACKGROUND TO THE METHODOLOGY

Decisions and the quantitative methods available to guide them may usefully be categorized in terms of four factors (Keeney and Raiffa, 1976, p. 27; Thrall, 1985). Table 2 shows the factors and the categorization which will be applied to the problem under discussion here. It is important to select a method well matched to the task at hand. The paper by Thrall gives an indication of some of the better known methods falling within each of the main categories, together with references to further sources of information.

On the basis of the outline descriptions given so far, it might well be objected that the categorization given in the final column of Table 2 is inaccurate. Surely, more than one decision maker will be involved? There must be some uncertainty about the accuracy of the numerical estimates of the impacts given in the matrix. The analyst's response can only be to agree. However, it is important to recognize that the assessment which the EDA is undertaking is a *modelling* exercise. Any quantitative model is an abstraction of the real world. It is a compromise between realism and tractability. Moreover, the output of the model is only a *guide* to policy choice, not a prescription. True, there is uncertainty. True, the final policy choice will probably reflect the views of many people, not just one, but often considerable insight about the relative strengths of the different strategies can be gained by the application of relatively straightforward methods which abstract from a lot of the finer detail. Moreover, although a more elaborate model might be built and calibrated which could in principle recognize more of the complexities of the real decision, the time, cost and manpower requirements for building such a model might prove prohibitive and, indeed, it might even come to be distrusted by the non-specialists involved in the decision, as an obscure 'black box' of impenetrable quantitative techniques handing out policy prescriptions on a basis with which they could not feel at ease. Good quantitative support to decision making is not necessarily the application of all the most sophisticated tools available.

Table 2
Categorization framework for decision problems

FACTORS	OPTIONS	CATEGORIZATION IN THIS CASE
The number of decision makers	One/many	One
The number of decision criteria	One/many	Many
The number of decisions	One/many	One
Is uncertainty formally recognized in the model?	Yes/no	No

A MULTIATTRIBUTE DECISION MODEL OF THE EDA PROBLEM

Multicriteria models which select between a finite set of alternatives of the type summarized in Table 1 (but do not directly construct new ones) are termed multiple attribute decision making (MADM) models. A comprehensive but straightforward survey is given in Hwang and Yoon (1981). Formally, the key inputs to MADM models of the type appropriate to this type of decision are:

(a) an $m \times n$ matrix of project 'scores' $\{X_{ij}\}$

where m = the number of projects (strategies) under consideration

  n = the number of separate evaluation criteria

  X_{ij} = the attribute score achieved by project i for evaluation criterion j

(b) a value function

$$V_i = f(X_{i1}, \ldots, X_{in}) \tag{1}$$

which maps any project characterized by a set of n attribute scores on to a one-dimensional value space, such that the higher the value of V_i, the more preferred is the project.

Before going further into the EDA's strategy choice problem, it is worth giving some attention to the two key components of the MADM model, the matrix of scores and the value function. Deriving the two of them is a major part of any multicriteria choice exercise and needs to be done with considerable care if the resulting evaluations are to reflect fairly the true values of the EDA. The main steps are: (a) structure the problem; (b) estimate the impacts (scores) of the alternative strategies; (c) determine the 'decision maker's' values and encapsulate them in an appropriate value function; before finally going on to (d) evaluate and compare the alternatives. Useful references which give more detail about what these steps involve are Keeney and Raiffa (1976), Keeney (1982, pp. 807–17) and Winterfeldt and Edwards (1986, Ch. 2, Sect. 2.1–2.4, and Chs. 7 and 8).

Structure the problem

In practice, identifying the appropriate problem structure is often *the* key determinant of a successful multicriteria evaluation. It requires the decision analyst to become properly conversant with what the real problem is; who is (are) the decision maker(s); what are their general objectives; by what

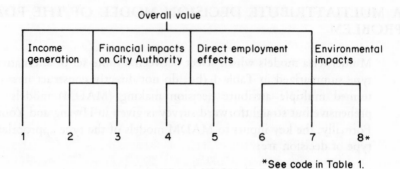

Figure 1
Value tree for the EDA problem

(measurable) attributes can the achievement of these objectives be assessed; and what are the alternative strategies that need to be evaluated? If answers to these questions are not properly established, there is a severe danger that a great deal of work will be put into providing a mathematically elegant answer to the *wrong* question, which will be of little or no practical value.

The end result of the structuring exercise should be a 'value tree', which for the EDA problem will be taken to be simply the one illustrated in Figure 1. Sometimes, as a way of helping to ensure that all relevant factors have been considered, it is useful in more complex problems to develop trees with several different levels (hierarchies); see Winterfeldt and Edwards (1986, pp. 36–45). Each final branch of the tree corresponds to an attribute relevant to assessing how well the available alternatives meet the decision maker's objectives in solving the current problem. The set of attributes should be assessed in terms of:

(a) *Completeness*. Are there any value-relevant aspects of the alternative strategies not present in the tree?

(b) *Measurability*. Can the alternatives' scores on each of the attributes be measured with the time/facilities/manpower available?

(c) *Redundancy*. Are any of the attributes effectively measuring the same thing?

(d) *Size*. Is the set of attributes small enough to allow effective analysis?

While completeness might encourage the listing of a very large number of attributes, the other three considerations argue powerfully for a more succinct representation of the decision maker's values.

Q1 Specify a problem of interest to you and a set of strategies that may be suitable to deal with that problem. Determine who the decision maker is, what his/her objectives are and what measurable attributes can be used to assess achievement of those objectives. Assess the set of attributes in terms of completeness, measurability, redundancy and size.

Estimate the impacts of the alternative strategies

The assessments of the impacts for the five strategies in the EDA problem are those given (earlier than they would normally be available) in Table 1. To assess how each alternative strategy would perform according to each of the eight selected attributes would normally in practice be another major task, perhaps involving the development of models of the local economic, transport, social systems, etc. Alternatively it might involve gathering the opinion of experts on these topics, if a formal quantitative modelling exercise was impracticable. For example, for the EDA problem there was not sufficient expertise or data available to quantify truly the changes in pollution associated with the different strategies, so an index of experts' subjective judgements is used. By no means could all of the understanding underlying Table 1 be incorporated into a formal quantitative representation of the problem. There is too much detail; the relationships between aspects of the understanding will often be obscure or vague. This is a further reason for not trying to build a highly sophisticated quantitative model. Understanding of this type is often better incorporated through the way in which a (relatively) simple evaluation model is used later in the decision process.

Determine the decision maker's values and develop a value function

This part of the overall process can be accomplished in several ways with varying levels of theoretical sophistication. For details of one of the more rigorous approaches, see Keeney and Raiffa (1976). Here, the approach that will be adopted is based on the simple multiattribute rating technique (SMART), as initially developed in Edwards (1977). SMART and similar techniques use a simple additive weighting (SAW) value function of the form:

$$V_i = \sum_{j=1}^{n} W_j V_j(X_{ij}) \tag{2}$$

By convention, normally $\sum W_j = 1$, since it is only the relative and not the absolute magnitudes of V_i that matter in determining the ranking of alternatives. The choice of a SAW value function does have what are, in principle, important implications for the nature of the decision maker's value structure and the manner in which scores on individual attributes contribute to the overall evaluation (Keeney and Raiffa, 1976, Ch. 3). In practice, it is often difficult to establish that the use of a SAW value function is strictly legitimate. However, the alternatives are relatively complex and there is a good deal of evidence to suggest that useful policy guidance is obtainable from simple linear models, even if the strict theoretical requirements for their use cannot be obeyed. Given the practicalities of data weakness, time constraints and political pressures on decisions, it is only for the most complex and important decisions that it usually makes sense to select a different type of value function.

The first step in specifying the parameters of the value function is to 'normalize' all the scores in the matrix such that, for each dimension separately, the best alternative scores 1 and the worst 0. For example, for impact 1 the normalization is

$$V_1(X_{i1}) = \frac{X_{i1} - 18}{50 - 18} \tag{3}$$

What is achieved through the normalization process is the construction of a series of single-dimensional linear value functions which can be used to evaluate the performance of any of the strategies in each dimension. Non-linear value functions are quite permissible, but the extent to which it is worth devoting the extra effort involved in specifying them varies from project to project (Winterfeldt and Edwards, 1986, Ch. 8).

Q2   Construct value functions for the remaining seven impacts, 2 to 8, and hence construct a rescaled impact matrix with all scores lying in the range 0–1. Note that while doing so, it is desirable to ensure that all scores are scaled such that 1 = 'best' and 0 = 'worst'. This may entail reversing the scale of some of the original scores by multiplying all of them through by −1 before normalizing them.

The main purpose of rescaling the project scores is to help in the process of generating reliable weights, the second major step in constructing the overall value function, V. It is important to be alert to the units in which the different impacts are measured when assessing the weights. There are several ways of calculating weights for a SAW model. Here, the ratio technique (Edwards, 1977) is used. (See Winterfeldt and Edwards, 1986, for a discussion of alternative ways of deriving weights and of the units of measurement.)

The ratio technique works as follows. First, taking careful note of the units of measurement, rank the (eight) impacts in increasing order of importance to the overall decision (ties are permissible). This ranking relates to unit changes in the *transformed* scales. For example, if heavy vehicle movements are rated least important and extra tax revenues next least important, this says that a one-unit increase in $V_7(X_{i7})$ (a fall of 700,000 vehicle movements per year on the original scale) is less important than a one-unit increase in $V_2(X_{i2})$ (an increase of £2m p.a. in tax and other revenues to the city).

Table 3
The EDA's ranking, rating and weighting of impacts

	IMPACT							
	1	2	3	4	5	6	7	8
Rank	=1	7	=4	=1	3	6	8	=4
Rate	150	20	50	150	100	30	10	50
Weight[a]	0.268	0.036	0.089	0.268	0.179	0.054	0.018	0.089

[a] Does not sum to 1 because of rounding error.

Next, assign the least important dimension a score of 10. Now, again being careful about units of measurement, rate the next least important dimension. If it is felt that a one-unit change in this variable is twice as important as a one-unit change in the first, the rating would be 20. Now proceed up the ranking list, comparing every set of ratios in the process and amending the ratings if necessary. Finally, all of the importance weights are summed and each is divided by the sum, yielding a set of eight W_j ($j = 1, \ldots, 8$) which sum to one.

Q3 Construct your own ranking of the eight attributes in the EDA problem. Rate them using the ratio method and finally derive your own set of W_j ($\sum W_j = 1$).

The EDA in fact derived the ranking, ratings and weights shown in Table 3. (Note that there is no reason why the EDA's assessment should be similar to yours.) There may be substantial but entirely legitimate differences in the relative importance attached by different groups to the different impacts. If, however, you feel at all uneasy about your weights, go back and check them, paying particular attention to the units of measurement.

Evaluate and compare the alternatives

All the necessary components for a MADM evaluation of the five strategies have now been developed. As a first stage in presenting an evaluation, it is always worthwhile to check for dominance. Alternative i dominates alternative k if $X_{ij} \geqslant X_{kj}$ for all j and $X_{ij} > X_{kj}$ for at least one j. The idea of dominance may be used either to identify one very good strategy or to demote and even dismiss from serious consideration a set of dominated, poor strategies. The next step is then to apply the derived value function, equation (2), to the individual strategies.

Q4 Check for dominance and then rank the five EDA strategies using both the EDA's weights, and your own, derived in Question 3.

It is most important to appreciate that in any substantial real-life decision making exercise, this ranking is *not* the end of the assessment process. Both the fact that the model is an approximate one and that the rankings above have not yet taken account of alternative viewpoints about weights have yet to be taken into account. Now is the time to start to *use* the model to help the EDA and the elected representatives to understand the relative strengths and weaknesses of the available alternatives.

One of the most effective ways of exploring the implications of a multicriteria evaluation exercise is through sensitivity analysis—exploring how the relative standing of the different projects changes as weights or scores are adjusted. Unfortunately, for most problems of realistic size, sensitivity analysis is a craft, not a science. There are simply too many possible combinations of weight and score changes to be able to explore and absorb into the analyst's understanding all the possible sensitivity tests that could be undertaken. Here, just two will be explained. A fuller discussion of sensitivity testing is available in Winterfeldt and Edwards (1986, Ch. 11).

On some occasions, there are potential doubts about the accuracy of the modelling which underlies the scores for a particular impact. In such a case it can be useful to re-run the evaluation exercise with an alternative set of scores and to examine the sensitivity of the dominance and ranking assessments to the change. Apart from simply displaying a revised set of strategy ranks, it is helpful to illustrate the changes graphically, perhaps using a simple bar chart.

Q5 Suppose one EDA official has serious doubts about the accuracy of the impact assessments for impacts 1 and 6 and suggests alternative scores for the five strategies as follows:

STRATEGY	IMPACT 1	IMPACT 6
A	45	20
B	35	10
C	18	10
D	23	35
E	32	10

Calculate the revised rankings which follow from adopting the new scores either separately or both simultaneously. Illustrate the effects graphically and discuss your findings.

The other impact dimension for sensitivity testing relates to the weights, W_j. It is common for different individuals or groups to have sharply contrasting views. Sensitivity to weight changes can be explored using the same methods as were used for impact score changes. Alternatively, the dependence of project ranking on a single weight, W_k, can be illustrated by plotting a graph of V_i as a function of W_k as follows.

Let W_k change in value through a parameter Δ. Since $\sum W_j = 1$, some adjustment of some or all the weights must occur if W_k changes. It is reasonable to assume that the adjustment is proportional, leading to

$$V_i = \sum_{\substack{j=1 \\ j \neq k}}^{n} \frac{W_j}{1+\Delta} V_j(X_{ij}) + \frac{W_k + \Delta}{1+\Delta} V_k(X_{ik})$$

When $\Delta = 0$, this is simply the original weighting system. By varying Δ from $-W_k$ upwards, the effect on overall ranking of changing W_k can be calculated and plotted on a graph for each strategy. Figure 2 illustrates the process for W_4. Sensitivity analysis of this type can often help defuse the argument that a particular impact has been underestimated, for example by illustrating that the overall ranking is insensitive to the weight it has been given.

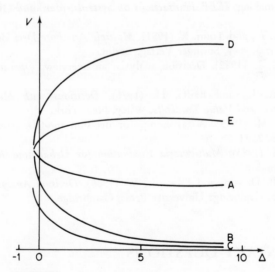

Figure 2
Sensitivity of V to changes in W_4

Q6 Illustrate diagrammatically the consequences of using your impact weights (Question 3) rather than the EDAs. In each case, illustrate diagrammatically the sensitivity of overall project ranking to the weight given to impact 5.

CONCLUSION

We have set out a number of tools which may be used to throw light on the strengths, weaknesses and overall rankings of a set of alternative strategies. All that remains is to put together the insights in a way that will help the responsible political representatives to make an informed choice. Often the main value of a multicriteria analysis is to force the parties involved to be explicit about the scores and weights and to avoid the tendency for proponents of a given strategy to concentrate on its strengths while overlooking its weaknesses. The final report should endeavour to reinforce this process.

Q7   In the light of Q1–6, write a report on the five proposed strategies to present to the city's elected representatives.

REFERENCES

Edwards, W. (1977). How to use multiattribute utility measurement for social decisionmaking, *IEEE Transactions on Systems, Man, and Cybernetics*, **SMC-7**, 326–40.

Hwang, C.-L., and Yoon, K. (1981). *Multiple Attribute Decision Making: A State of the Art Survey*, Springer-Verlag, Berlin.

Keeney, R. L. (1982). Decision analysis: an overview, *Operations Research*, **30**, 803–38.

Keeney, R. L., and Raiffa, H. (1976). *Decisions with Multiple Objectives: Preferences and Value Tradeoffs*, Wiley, New York.

Thrall, R. M. (1985). A taxonomy for decision models, *Annals of Operations Research*, **2**, 23–7.

Voogd, H. (1983). *Multicriteria Evaluation for Urban and Regional Planning*, Pion, London.

Winterfeldt, D. von, and Edwards, W. (1986). *Decision Analysis and Behavioral Research*, Cambridge University Press, Cambridge.

SUPPLEMENTARY QUESTIONS

1. Establishing weights is often difficult in multicriteria evaluation exercises. An alternative (or supplementary) procedure which can sometimes be used is to employ multiple regression to estimate the weights, based on a record of the input scores of projects analysed in the past and an estimate of each project's overall value once implemented. Suppose that the following information is available :

ESTIMATED PROJECT VALUE V	IMPACT SCORES							
	1	2	3	4	5	6	7	8
0.517	0.170	0.220	0.398	0.843	0.264	0.005	0.089	0.774
0.492	0.012	0.373	0.214	0.691	0.009	0.950	0.147	0.132
0.487	0.877	0.423	0.116	0.092	0.370	0.821	0.427	0.727
0.407	0.362	0.517	0.216	0.842	0.311	0.074	0.512	0.021
0.381	0.117	0.217	0.631	0.125	0.111	0.853	0.571	0.317
0.411	0.625	0.361	0.337	0.289	0.277	0.591	0.463	0.816
0.389	0.224	0.847	0.019	0.387	0.584	0.211	0.673	0.104
0.217	0.732	0.669	0.951	0.124	0.773	0.218	0.912	0.487
0.410	0.371	0.462	0.801	0.103	0.321	0.557	0.702	0.130
0.227	0.611	0.501	0.179	0.399	0.487	0.234	0.510	0.173
0.631	0.092	0.375	0.714	0.822	0.637	0.518	0.157	0.264
0.442	0.287	0.423	0.616	0.358	0.272	0.918	0.247	0.281
0.471	0.129	0.442	0.801	0.628	0.238	0.402	0.219	0.237
0.690	0.752	0.201	0.315	0.842	0.430	0.787	0.231	0.529
0.137	0.057	0.831	0.423	0.103	0.211	0.101	0.168	0.057
0.503	0.427	0.713	0.382	0.366	0.517	0.713	0.131	0.218
0.101	0.003	0.257	0.416	0.089	0.214	0.113	0.414	0.281
0.591	0.373	0.518	0.316	0.274	0.589	0.922	0.371	0.597
0.635	0.801	0.235	0.115	0.762	0.501	0.748	0.444	0.397
0.321	0.417	0.339	0.847	0.419	0.083	0.768	0.395	0.981

Use the above data to estimate W_j $(j = 1, \ldots, 8)$ and hence assess the ranking of the five projects. What are the advantages and disadvantages of using a regression approach to establish attribute weights?

2. An alternative to establishing precise numerical values for the weights is to specify ranges within which the true weight must lie. For example,

$$W_1 \geqslant W_3 \qquad W_5 \geqslant W_6$$

$$W_3 \geqslant W_7 \qquad W_1 \geqslant 0.1$$

$$W_1 + W_4 \leqslant 0.5 \qquad W_6 \geqslant W_2$$

$$W_8 \geqslant 0.05 \qquad W_5 \leqslant W_1 + W_4$$

Although it is now impossible to calculate a precise V_i, alternative strategies can be compared on the basis of the maximum and minimum weighted scores they can achieve. It is also sometimes possible to establish dominance; that is given the ranges specified for the W_j, it is impossible for one particular project to have weighted scores better than/worse than some other(s).

Given the constraints set out above, use a linear programming package to rank the five EDA strategies in terms of their maximum and their minimum attainable weighted scores. By looking at every possible pair of sets of differences in scores $D_j^{ik} = V_j(X_{ij}) - V_j(X_{k_j})$ $(j = 1, \ldots, 8)$, check whether dominance exists between any pairs of strategies. (Hint: if $\min \sum W_j D_j^{ik} > 0$, then i dominates k; if $\max \sum W_j D_j^{ik} < 0$, then k dominates i.) Hence, comment on the relative strength of the five strategies. (See Kmietowicz and Pearman, 1984.)

3. What are the advantages and disadvantages of the MADM approach to project evaluation and ranking as compared to the conventional economics-based approach of cost–benefit analysis? What alternatives are there? (See Pearce and Nash, 1981, Sec. 2.4; Watson, 1981.)

4. Either individually, or as a member of a small group, choose an evaluation and choice problem that concerns you and examine the options which are available, using the methods set out in the assignment. Pay particular attention to problem structure, estimation of impact weights, sensitivity analysis and presentation of your results.

FURTHER REFERENCES

Kmietowicz, Z. W., and Pearman, A. D. (1984). Decision theory, linear partial information and statistical dominance, *Omega*, **12**, 391–9.

Pearce, D. W., and Nash, C. A. (1981). *The Social Appraisal of Projects*, Macmillan, London.

Watson, S. R. (1981). Decision analysis as a replacement for cost/benefit analysis, *European Journal of Operational Research*, **7**, 242–8.

Assignments in Applied Statistics
Edited by S. Conrad

A Decision Tree for the Selection of Tenderers

J. Q. Smith

Department of Statistics, University of Warwick

OUTLINE

A manager must decide which firms to invite to tender for two maintenance contracts. The optimal tendering policy is to be determined from a decision tree. Certain necessary assumptions can be represented by an influence diagram, to aid understanding. The possibility of an analytical solution and the generalization of the problem, involving the pruning of what becomes a 'bushy tree', are considered in the supplementary questions.

KEYWORDS

decision trees, modelling assumptions, minimization of expected cost, optimal policy, conditional probability, extensive form analysis, expected value of sampling information (EVSI), influence diagrams, general solution, bushy trees, stochastic control

A large company is often required to employ contractors to periodically undertake structural repairs for them. When large amounts of money are involved in such a transaction it is usual for higher management to require middle management to put that work out 'to tender' to firms of contractors. This policy is supposed to act as a crude safeguard against bribery and corruption in middle management.

To be more explicit, a job specification will be drawn up by an expert within the company. The middle manager will then select a pre-stated number, n, from m possible contracting firms who will then be invited to quote a price for the work. The number, n, of firms that will be tendering will vary with the type of work. For moderately expensive work, the middle manager will perhaps invite tenders from only two firms; for more costly work, the manager will perhaps invite tenders from as many as six firms.

Each of the n contractors will have two decisions to make:

(a) whether to tender at all and

(b) if tendering, what price to quote for the work.

When all tenders have been received, unless there are strong reasons for doing otherwise, the middle manager must employ the firm who submits the lowest tender. Essentially, therefore, your client's only real decision is to choose which n out of the m possible contractors will be asked to tender for any job.

In making this decision, one of the main objectives of the manager will be to ensure that the policy of selecting which companies to invite to tender contains at least one low tender so, at least in the long run, work is done as cheaply as possible. Other objectives which might also come into play are:

(a) to ensure than any tendering company has sufficient financial backing for the work;

(b) to ensure that the work will be done at least adequately.

Contracting firms tend to be very unpredictable in the amount of profit they expect to make from one piece of work for which they are tendering. A common reason for a firm submitting a very low tender is that it badly needs the work. However, other factors may come into play. Some firms may be geared up to do precisely the work that is required so that they are able to quote a low price whilst still maintaining a reasonable profit margin. Other firms, although not needing work now, may submit a fairly low tender now so that they will be asked to tender for future similar contracts when they do need work.

Once the tenders have been received, the only legitimate information an unsuccessful tenderer can learn is who obtained the tender, although sometimes they will be told informally whether their price was close to that of the successful tender. However, the exact price quoted in the successful tender (or any other tender) will not be divulged to any other firm.

THE MAINTENANCE CONTRACTS UNDER CONSIDERATION

You are asked by a middle manager to advise him about his best course of action in respect of two similar jobs which he will need to put out to tender, one three months after the other. Because of the time lag between jobs, he will have received tenders for the first job before he needs to select which firms to ask to tender for the second.

The two jobs he requires happen to be highly specialized and there are only three firms F_1, F_2, F_3 (i.e. $m = 3$) who would be able to submit a tender. Because the cost of each job is relatively small, he is required by higher management to invite just two ($n = 2$) firms to submit a tender. Each piece of work will be tendered for separately so that your client is not constrained to choose the same two firms to submit tenders for both jobs.

Your client has checked each of the three firms and found them to be on a sound financial footing. Because the work you offer is potentially very lucrative to any firm winning the work, you believe that any firm who is asked to tender will certainly choose to quote a price. Your client has told you that his primary objective is to choose those two sets of tenderers so as to minimize the expected cost of the work as aggregated over the two maintenance contracts.

Firm F_3 is large and its tenders are very predictable. In particular their stated price will not depend upon their need for work, which is always fairly constant anyway. On the other hand, the expected tenders from F_1 and F_2 depend heavily on whether they need the work [denoted by the event $S(t)$] or do not need the work [denoted by the event $\bar{S}(t)$] [for job t ($t = 1,2$), where the index t labels the job which is being put out to tender]. The expected tenders for each job based on the previous tenders of the

Table 1
Expected tender given need [$S(1)$] or no need [$\bar{S}(1)$] of work (in units of £10,000)

NEED OF WORK	FIRM		
	F_1	F_2	F_3
$S(1)$	13	12	15
$\bar{S}(1)$	20	18	15

Table 2
Probabilities that firms F_1 and F_2 need work at the time of the first job

NEED OF WORK	FIRM	
	F_1	F_2
$S(1)$	0.6	0.5
$\bar{S}(1)$	0.4	0.5

Table 3
Probabilities that firms F_1 and F_2 need or do not need work at the time of the second job, given that they need/do not need work at the time of the first job

	$S(2)$	$\bar{S}(2)$
$S(1)$	0.8	0.2
$\bar{S}(1)$	0.2	0.8

three firms are given in Table 1. The probabilities that the smaller firms F_1 and F_2 need work at the time of the first job are given in Table 2.

If a firm needs work at the time of the first job it is very probable that it will need work at the time of the second. The probabilities that firms F_1 or F_2 need work at the time of the second job given that they need, or have no need of, work on the first is given in Table 3.

ASSUMPTIONS

Acting on behalf of your client you are prepared to make the following additional modelling assumptions in order to obtain an answer to his problem:

(a) Given the need, or otherwise, for work by any firm, the tender of the firm will not deviate by more than £5,000 of its expected value as stated in Table 1.

(b) Firm F_1's need for work for job 1 is independent of F_2's need for work for job 1 or job 2, and conversely F_2's need for work at the time of job 2 is independent of F_1's need for work at the time of job 1 or job 2.

(c) Let $X_i(t)$ denote F_i's tender for job t $(i = 1,2,3; t = 1,2)$. Let $S_i(t)$ denote the indicator random variable that F_i needs work at the time of job t $(i = 1,2; t = 1,2)$.

   (i) Suppose you know whether or not F_1 needs work at the time of the first job. Then no additional information about F_1's tender can be deduced by knowing any other variable in the model. (In probabilistic terms we are saying that conditional on $S_1(1)$, $X_1(1)$ is independent of $[X_1(2), X_2(1), X_2(2), S_1(2), S_2(1), S_2(2), X_3(1), X_3(2)]$.)

(ii) Similarly, if you know whether or not F_2 needs work at the time of the first job, then no additional information about F_2's tender can be deduced by knowing any other variable in the model. (In probabilistic terms, conditional on $S_2(1)$, $X_2(1)$ is independent of $[X_2(2), X_1(1), X_1(2), S_2(2), S_1(1), S_2(2), X_3(1), X_3(2)]$.)

(iii) Again, if you knew whether or not F_1 needed work at the time of the second job, then no other quantity need be used in your assessment of F_1's probable second tender. (Probabilistically, conditional on $S_2(2)$, $X_2(1)$ is independent of $[X_1(1), X_1(2), S_1(1), S_1(2), S_2(2), X_3(1), X_3(2)]$.)

(iv) Also, if you knew whether or not F_2 needed work at the time of the second job, then no other quantity need be used in your assessment of F_2's probable second tender. (Probabilistically, conditional on $S_2(2)$, $X_2(1)$ is independent of $[X_1(1), X_1(2), X_2(2), S_1(1), S_1(2), S_2(1)]$.)

(v) Both $X_3(1)$ and $X_3(2)$ are independent of any other variable. (Probabilistically, $X_3(1)$ and $X_3(2)$ are independent and $[X_3(1), X_3(2)]$ is independent of $[X_1(1), X_1(2), X_2(1), X_2(2), S_1(1), S_1(2), S_2(1), S_2(2)]$.)

Q1 Your client has three different decisions associated with each piece of work. Represent the problem as a decision tree, calculating:

(a) all necessary probabilities of chance forks needed on the tree;
(b) all necessary terminal expected payoffs associated with the tree.

(Decision trees are discussed in, for example, Raiffa, 1968, Ch. 2; Moore and Thomas, 1976, Ch. 4; Buchanan, 1982, Ch. 6; Lindley, 1985, Ch. 8; and Smith, 1988, Ch. 2.)

By using an extensive form analysis of the problem (i.e. by drawing a decision tree), advise the client both on the best choice of the two tenderers for the first job and the best choice of the two tenderers for the second job depending on the first. What is the expected cost under this optimal choice of action?

Q2 Now your client has seen the result of your calculations he is interested in how your simplifying assumptions might have influenced the analysis. Explain why assumption (a) was necessary to keep the analysis simple. Also discuss when he might expect assumption (b) not to hold and why assumption (c) might be violated under certain circumstances.

Q3   Your client now tells you that in the past he has always chosen the two firms expected to give the lowest tender on a particular job. Give two reasons why this is not an optimal way to behave here if the intention is to minimize the expected total cost of the two jobs.

REFERENCES

Buchanan, J. T. (1982). *Discrete and Dynamic Decision Analysis*, Wiley, Chichester.

Lindley, D. V. (1985). *Making Decisions*, Wiley, Chichester.

Moore, P. G., and Thomas, H. (1976). *The Anatomy of Decisions*, Penguin, Harmondsworth, Middlesex.

Raiffa, H. (1968). *Decision Analysis*, Addison-Wesley, Reading, Massachusetts.

Smith, J. Q. (1988). *Decision Analysis: A Bayesian Approach* , Chapman and Hall, London.

SUPPLEMENTARY QUESTIONS

1. Your client is able to hire, on behalf of the company, a private investigator. This investigator would be able to determine, for certain, whether or not firms F_1 and F_2 need work just before your client needs to choose the two tendering firms for both pieces of work that need doing. Find the expected value of this information. Putting ethical considerations aside, how much should the client be prepared to pay for such information?

2. The (conditional) independence assumptions given in assumption (c) are difficult to understand at first sight. Draw an influence diagram to represent the relationship between the various variables given in this set of assumptions. Hence or otherwise draw an influence diagram representing the problem as a whole. (Influence diagrams are discussed in, for example, Bunn, 1984, Ch. 11; Schachter, 1986; and Smith, 1988, Ch. 5.)

3. In general, finding solutions to tendering problems involves a considerable amount of computation and needs to be performed on a computer. However, if we are prepared to make some convenient distributional assumptions then it is possible to obtain some more general results that give insight into why we obtain the solutions we do.

   Let us suppose that your client is interested in minimizing costs over a single piece of work. The client needs to choose the n firms which will be asked to tender from a set $\{F_1, F_2, \ldots, F_m\}$ where the tender price X_i of F_i is distributed $E(a, p_i, b)$, $1 \leqslant i \leqslant m$, where $E(a, p, b)$ has the distribution function $G(x)$ given by

$$G(x) = \begin{cases} 0, & x < a \\ 1 - p, & x = a \\ (1 - p) \exp[-b(x - a)], & x > a \end{cases}$$

(If a firm has its tender distributed as above then it could be said that with probability $1 - p$ it will offer to do the work at a minimum possible price a and has a probability of offering a tender higher than a. These higher tenders will have an exponential distribution with mean parameter b^{-1} and location shift a. Assume that all prices are quoted independently of each other.)

(a) Show that the minimum tender of any collection of n firms has a distribution $E(a, p, b)$ and relate the parameters of this minimum to those of the tender price of each of the firms in your collection.

(b) Your client needs to choose between firms F_n and F_n^* to add to a collection of $n - 1$ other firms $\{F_1, \ldots, F_{n-1}\}$ which will be asked to tender. The expected tenders X_n and X_n^* respectively are the same. If all firms have tender prices of the form above, show that your client should ask F_n to tender in preference to F_n^* if and only if the variance of X_n is at least as great as the variance of X_n^*.

(The distribution of the minimum of a set of independent variables can be found in most introductory texts in mathematical statistics, and also in Gibbons, 1971.)

4. In this assessment we have assumed that the maintenance contracts would only be offered twice and that two firms from three are to be chosen at each stage. It is far more usual for there to be a much longer series of jobs on offer and the pool of available firms to be much larger. Obviously this would produce a much more bushy tree. Discuss some of the ways in which such a tree could be pruned down to a manageable size and recommend a decision rule which, if not necessarily optimal, is still likely to be close to optimal.

5. Suppose that you are now required to advise on a company policy that instructs managers how to act when they have to choose which two firms from three tender for maintenance contracts which occur on a regular basis. Show that this is possible provided that you are prepared to use some form of temporal discounting. Write this down as a stochastic control problem and indicate how you might attempt to solve it.

(The tendering problem above, together with the problems inherent in its analysis, is in fact a simple example of a stochastic control problem. These types of problems are dealt with in detail at a more advanced level in, for example, DeGroot, 1970, or Whittle, 1983.)

FURTHER REFERENCES

Bunn, D. W. (1984). *Applied Decision Analysis*, McGraw-Hill, New York.
DeGroot, M. H. (1970). *Optimal Statistical Decisions*, McGraw-Hill, New York.
Gibbons, J. D. (1971). *Nonparametric Statistical Inference*, McGraw-Hill, New York.
Schachter, R. D. (1986). Evaluating influence diagrams, *Operations Research*, **34**, 871–82.
Whittle, P. (1983). *Optimisation Over Time*, Volume 2, Wiley, New York.

Assignments in Applied Statistics
Edited by S. Conrad
© 1989 John Wiley & Sons Ltd

Sample Information in Marketing a New Product

Z. W. Kmietowicz

School of Business & Economic Studies, University of Leeds

OUTLINE

A classical decision problem of whether or not to market a new product is considered. The decision maker initially relies on readily available information. The advisability of acquiring additional information by employing a market research company and conducting a sample survey is then considered. The determination of the optimal strategy requires the updating of probabilities through Bayes' theorem, and backward induction involving the pruning of inoperative branches of the decision tree. The decision problem is then reformulated by expressing profit as a linear function of a random variable. Decision making with incomplete information is also considered. This involves the calculation of minimum and maximum expected profits of strategies when the probabilities of states of nature are ranked, and the use of maximin and maximax criteria for strategy determination. Statistical dominance and sensitivity testing are also considered.

KEYWORDS

Profit matrix, expected value, utility, subjective probability, expected value of perfect information (EVPI), decision trees, conditional probability, Bayes' theorem, updating of probabilities, backward induction, optimum sample size, profit functions, Unit Normal loss integral, incomplete information, ranking of probabilities

A large company producing soft drinks has developed a new product and is considering whether it should be marketed. The success or failure of the product depends on its reception by the general public and on the resulting sales. Both the range and the probability distribution of sales are usually difficult to specify exactly and the analysis is simplified by restricting the different types of reception to: good (G), moderately good (M) and poor (P). (The method of dealing with a fully specified probability distribution of sales is discussed on page 47.)

Using its experience of previous launches, management estimates that the reception of the new product will be 'good' with probability 0.2, 'moderately good' with probability 0.3 and 'poor' with probability 0.5. Management also estimates that if the reception is good, the company will make a profit of £200m; if moderately good, a profit of £20m; and if poor, a loss of £100m. Development costs of the new product are estimated to be £5m. (These costs have been taken into account when estimating the profits and loss mentioned above.)

Q1 Assuming the company has to make a decision quickly, relying only on information available to it, specify the strategies (S_i), the states of nature (N_j) and their probabilities (p_j), and the profit matrix with elements (X_{ij}) representing profit or loss arising when strategy i is chosen and state of nature j occurs. Using the maximum expected profit criterion, select the best course of action open to the company.

Q2 Represent the decision problem facing the company in the form of a decision tree. Insert the information available to the company on the decision tree, distinguishing carefully between decision and event nodes. Evaluate expected profit at the event nodes and select the best course of action using maximum expected profit as the criterion.

Q3 Comment on the implications of the course of action chosen in Question 1 and compare it with the result obtained in Question 2.

Q4 Does it make any difference whether the decision facing the company is unique (i.e. unlikely to be repeated soon) or routine (i.e. similar to other decisions made regularly)? When does the minimum profit or maximum loss of the selected strategy present special difficulties to the company?

Q5 Does the monetary profit or loss necessarily represent the attractiveness to the company of a strategy operating under a particular state of nature? If it does not, how can money profits or losses be converted into decision maker's subjective utilities? How can the optimum strategy be selected in this case?

Q6 How are the probabilities of the states of nature obtained? Are they likely to be accurate? If not, can you suggest an alternative way of stating any information the decision maker may have about them? If an alternative estimate of the probabilities is $p_1 = 0.2$, $p_2 = 0.4$ and $p_3 = 0.4$, would the strategy selected in Question 1 still be optimal?

Q7   Management also suspects that the estimated profit when the new product is well received may be exaggerated. If the estimate is reduced from £200m to £150m and the estimated probabilities of the states of nature are those used in Question 6, what course of action would be optimal now?

(See Miller and Starr, 1969, Chs. 4 and 5; Moore, 1972, Chs. 5 and 9; Moore and Thomas, 1976, Chs. 2–4; Kmietowicz and Pearman, 1981, Ch. 2; Bunn, 1982; French, 1986, Ch. 2.)

USE OF A MARKET RESEARCH COMPANY

As the analysis performed in Questions 1 to 7 is inconclusive, management considers a third course of action, that is employment of a reputable market research company (MRC) in order to obtain more reliable information about the likely reception of the new product by the general public. Management knows that the cost of hiring the services of the MRC will be £2m and that its forecasts, although fairly reliable, are by no means perfect. The MRC is accustomed to assessing the prospects of a new product as very favourable (VF), favourable (F) and unfavourable (UF). Previous dealings with the MRC suggest that joint probabilities of state of nature and forecast are as shown in Table 1.

Table 1
Joint probabilities of state of nature and type of forecast

STATE OF NATURE	FORECAST BY THE MRC		
	VF	F	UF
G	0.10	0.07	0.03
M	0.15	0.10	0.05
P	0.10	0.18	0.22

For example, the probability that the reception of the new product will be good and the forecast will be favourable is $p(G \cap F) = 0.07$.

Q8   Amend the decision tree obtained in Question 2 to incorporate the possibility of employing the MRC as the third strategy open to management. Distinguish carefully between decision and event nodes.

Assuming that the MRC can predict the reception of the new product by the general public with perfect accuracy, calculate the expected payoff of perfect information (EPPI) and hence the expected value of perfect information (EVPI). Does the EVPI indicate that employment of the MRC may be advantageous?

Q9 Use the information given in Table 1 to calculate the probabilities of the three types of forecasts made by the MRC and the conditional probabilities of states of nature given the MRC forecasts, and insert them on the decision tree.

Using backward induction, evaluate expected profits at event nodes, prune the inoperative branches of the decision tree and establish whether employing the MRC is preferable to the course of action selected in Question 1. If the MRC is employed, indicate clearly what action management should take in response to the different forecasts it could make.

The joint probabilities shown in Table 1 are usually difficult to estimate. It is more likely that management will have some information about the likely accuracy of forecasts made by the MRC. This information may come from previous dealings with the company or from its reputation. Assume that the conditional probabilities given in Table 2 represent a fair summary of the accuracy of the forecasts made by the MRC.

Q10 Using the probabilities given in Table 2, where, for example, $p(F/G) = 7/20$, and the unconditional probabilities of the states of nature used in Question 1, calculate all the probabilities needed for the expanded decision tree. Are the probabilities the same as those obtained in Question 9? What course of action should management choose now?

Table 2
Probabilities of type of forecast conditional on state of nature

STATE OF NATURE	FORECAST BY THE MRC		
	VF	F	UF
G	1/2	7/20	3/20
M	1/2	1/3	1/6
P	1/5	9/25	11/25

Q11 If the only probabilities available to management were the unconditional probabilities of the states of nature given on page 42 and the conditional probabilities shown in Table 3, where, for example,

Table 3
Probabilities of state of nature conditional on type of forecast

STATE OF NATURE	FORECAST BY THE MRC		
	VF	F	UF
G	2/7	7/35	3/30
M	3/7	10/35	5/30
P	2/7	18/35	22/30

$p(G/F) = 7/35$, would it be possible to obtain all the probabilities required for the expanded decision tree? If your answer is yes, show how this can be done; if not, explain why not and state what additional information is needed. If additional information is required, can it be obtained from Table 1? If it can, show how the required probabilities can be calculated.

Q12   Compare the prior probabilities of states of nature employed in Question 1 with the posterior probabilities of states of nature obtained in Question 9, that is the probabilities of the states of nature conditional on the MRC forecasts. In what sense can the posterior probabilities be said to be revised prior probabilities? What role does Bayes' theorem play in the revision?

Q13   If one of the directors involved in the decision making process believes that more accurate estimates of the probabilities of the MRC forecasts being VF and F, when prospects for the new product are in fact good, are 0.40 and 0.25 respectively, rather than 1/2 and 7/20 as assumed in Table 2, calculate the required probabilities and employ backward induction again to identify the optimum strategy. Compare your results with those obtained in Question 10.

(See Miller and Starr, 1969, Ch. 9; Moore, 1972, Chs. 6–8; Moore and Thomas, 1976, Chs. 5 and 6; Buchanan, 1982, Chs. 5 and 6.)

USE OF SAMPLE INFORMATION

Another member of the decision team, who believes that the fee charged by the MRC is too high, proposes an extensive trial with a sample of people as an alternative. This proposal requires a redefinition of the states of nature in terms of the proportion of the general public likely to buy the new product in the long run. If the proportion is 0.3, the company will make a profit of £200m; if it is 0.2, it will make a profit of £20m; and if it is 0.1, it will make a loss of £100m. The probabilities of the three states of nature mentioned above are still 0.2, 0.3 and 0.5 respectively. Instead of engaging the MRC, the management of the soft drinks company decides to take a random sample of people chosen from the whole nation and to conduct extensive trials with them lasting several months. At the end of the trials the customers have to state whether they are likely to buy the product or not. The total cost of such trials is estimated to be £0.5m per person.

Q14   Amend the decision tree obtained in Question 2 to include the new strategy of taking a random sample of two people and subjecting them to intensive trials. Distinguish carefully between event and decision nodes. Obtain conditional probabilities of states of nature given sample outcomes, and unconditional probabilities of sample outcomes, and insert them in the decision tree.
       Use backward induction again to establish whether the new strategy is preferable to the strategies available in Question 2. Is the new strategy preferable to employing the MRC?

Q15   Amend the decision tree obtained in Question 14 by including a new strategy of taking a random sample of three people and subjecting them to extensive trials. Obtain the probabilities required for the branches of the decision tree describing the options presented by the new strategy. Use backward induction again to evaluate the attractiveness of the new strategy. Is the new strategy preferable to employing the MRC? Is the new strategy preferable to employing a random sample of two people?

Q16   Without performing any further calculations, explain how optimum sample size can be determined.

(Revising or updating probabilities through Bayes' theorem for discrete and continuous variables is discussed in Schlaifer, 1969; Moore, 1972, Chs. 6–8; Moore and Thomas, 1976, Chs. 5 and 6; Buchanan, 1982, Chs. 5 and 6; Bunn, 1982; French, 1986, Chs. 6 and 7.)

DECISION PROBLEMS WHERE PROFIT IS A LINEAR FUNCTION OF A RANDOM VARIABLE

It is assumed here that the company producing soft drinks is considering launching one of three new products. It is unable to launch more than one product because of its poor financial position. Development costs of the new products are similar, but product A is likely to do well if national disposable income during the next two years increases, product B is expected to be popular if national disposable income declines and product C is likely to do best if national disposable income remains stationary. The company estimates that the profit functions of the three products are:

$$f(A, \theta) = -16 + 2\theta$$
$$f(B, \theta) = 6 - \theta$$
$$f(C, \theta) = 4$$

where θ is the percentage change in national disposable income during the next two years and profits are measured in millions of pounds.

Q17   Graph the profit functions of the three products and determine the range of values of θ for which each of the products is optimal, that is find the breakeven points. If the probability distribution of θ is assumed to be Normal with mean 5 per cent and standard deviation 4 per cent, which product should the company launch? Obtain the expected value of perfect information (EVPI) and explain what use can be made of it. (Note: first, define the opportunity loss function for different values of θ and then use it to find the EVPI.) (Table 29 of the Unit Normal loss integral in Kmietowicz and Yannoulis, 1988, may be useful.)

Q18   If another new product (D) with the profit function

$$f(D, \theta) = 3.2 + 0.4\theta$$

is included in the decision problem, which product should be selected now? What is the EVPI for this problem? Compare your result with that obtained in Question 17.

(See Schlaifer, 1969; Moore, 1972, Chs. 5, 6 and 11; Buchanan, 1982, Chs. 5 and 6; Bunn, 1982; French, 1986, Chs. 6 and 7.)

DECISION PROBLEMS WITH INCOMPLETE INFORMATION

In most decision problems it is assumed that the probabilities of future states of nature can be specified exactly by the decision maker. This is difficult, if not impossible, to do in many business and management problems. Consider an investment problem where the states of nature represent economic conditions likely to arise in the future, for example

Table 4
Profit matrix

| | STATE OF NATURE | | |
STRATEGY	N_1	N_2	N_3
S_1	10	4	−4
S_2	6	5	3
S_3	4	5	8

favourable, unfavourable and unchanged. In this case the decision maker is unlikely to be able to specify accurately the probabilities of future states of nature. If pressed by the decision analyst, the decision maker may supply a set of exact probabilities, but will be aware of the tentative nature of such probabilities. The decision maker may prefer to give incomplete information about the probabilities in a way which is closer to reality. Here it is assumed that the decision maker is able to rank the probabilities in order of their magnitude, but may feel unable to specify them exactly. (See Kmietowicz and Pearman, 1981, Ch. 3.)

As an illustration, consider a decision problem similar to the one discussed on page 42. The company is planning to launch one of three new products submitted to it by the product development department. The estimates of profitability of the three products take into account development, capital and marketing costs, as well as different pricing policies which will be pursued under different economic conditions which may arise in the future. Three future states of nature are considered: expanding economy (N_1), static economy (N_2) and declining economy (N_3). The strategies available to the decision maker correspond to the launching of one of the three products. The estimated profits (in millions of pounds) for the three products under the three states of nature are shown in Table 4.

Q19 If the exact probabilities of the states of nature are assumed to be $p_1 = 11/18$, $p_2 = 5/18$ and $p_3 = 2/18$, which product should be selected if the decision maker wishes to maximize expected profit?

Q20 Assume now that the decision maker is unable to specify the exact probabilities of the states of nature, but is only able to rank them, that is, $p_1 \geq p_2 \geq p_3$. Check whether any of the strategies dominates another in the sense that it yields a higher expected profit than the dominated strategy for all probability distributions satisfying the ranking constraints. (This type of dominance is called *statistical dominance* and should be distinguished from *mathematical dominance* which occurs when profits of one strategy are larger than profits of another strategy for all states of nature.)

Q21 Given the above ranking of the probabilities, formulate a linear program which would enable the decision analyst to calculate the minimum and maximum expected profits for a particular strategy. Can you find a *general* solution to the problem? Interpret your results.

Q22 Given the ranking of probabilities specified in Question 20, obtain the minimum and maximum expected profits for each strategy. If the decision maker uses the maximin criterion applied to the minimum expected profit of the different strategies, which product should be chosen? If the maximax criterion applied to the maximum expected profit of the different strategies is favoured, which product should be chosen? If the maximin and maximax criteria are applied to the minimum and maximum profits of the different strategies would the choice of strategies alter? Would you expect the maximin criterion applied to the minimum expected profit of the different strategies always to be preferable to the maximin criterion applied to the minimum profit of the different strategies?

Q23 The problem of incomplete information about the probabilities of the states of nature can also arise in situations like the ones discussed in Questions 1 and 12. Reconsider the problem discussed in Question 1, if it is assumed that management cannot specify exactly the probabilities of the states of nature, but can rank them as follows: $p(G) \leqslant p(M) \leqslant p(P)$. If the conditional probabilities required in Question 9 are not known exactly but can be ranked, suggest a procedure for identifying the optimum strategy. When information about the probabilities of the states of nature is incomplete, can the calculation of minimum and maximum expected profits for the optimal strategy be viewed as sensitivity testing? Does such sensitivity analysis have any advantages as compared with traditional sensitivity testing of the optimal strategy?

Q24 In Questions 20 and 22 it was assumed that the decision maker could rank the probabilities of the states of nature. Can you suggest other linear constraints which could be usefully placed on the probabilities in order to describe more accurately the information available to the decision maker? Can you formulate linear programs which would enable you to find minimum and maximum expected profits for strategies in this case? Can you find *general* solutions to such programmes? What uses could be made of these results?

Q25 If the decision maker is able only to rank the probabilities of the states of nature, but the decision analyst, brought up on traditional methods of decision making under conditions of risk, requires uniquely specified probabilities, can you suggest a method of obtaining such probabilities from the ranking constraint?

(Further discussion of decision making with incomplete information may be found in Kmietowicz and Pearman, 1981, 1984; Kofler, Kmietowicz and Pearman, 1984.)

REFERENCES

Bunn, D. W. (1982). *Analysis for Optimal Decisions*, Wiley, Chichester.
Buchanan, J. T. (1982). *Discrete and Dynamic Decision Analysis*, Wiley, Chichester.
French, S. (1986). *Decision Theory*, Ellis Horwood, Chichester.
Kmietowicz, Z. W., and Pearman, A. D. (1981). *Decision Theory and Incomplete Knowledge*, Gower Press, Aldershot.
Kmietowicz, C. W., and Pearman, A. D. (1984). Decision theory, linear partial information and statistical dominance, *Omega*, 12, 391-9.
Kmietowicz, Z. W., and Yannoulis, Y. (1988). *Statistical Tables for Economic, Business and Social Studies*, Longman, London.
Kofler, E., Kmietowicz, Z. W., and Pearman, A. D. (1984). Decision making with linear partial information (L.P.I.), *Journal of the Operational Research Society*, 35, 1079-90.
Miller, D. W., and Starr, M. K. (1969). *Executive Decisions and Operations Research*, Prentice-Hall, Englewood Cliffs, New Jersey.
Moore, P. G. (1972). *Risk in Business Decision*, Longman, London.
Moore, P. G., and Thomas, H. (1976). *The Anatomy of Decisions*, Penguin, Harmondsworth, Middlesex.
Schlaifer, R. (1969). *Analysis of Decisions under Uncertainty*, McGraw-Hill, New York.

SUPPLEMENTARY QUESTIONS

1. Obtain the regret or opportunity loss matrix from Table 4. Use it to select a product when the probabilities of the states of nature given in Question 19 are assumed to apply and the decision maker wishes to minimize expected regret. Compare your answer with that given to Question 19. Discuss the implications of the comparison.
2. Using the regret matrix obtained in the previous question and assuming that the probabilities cannot be specified accurately but can be ranked as in Question 20, obtain minimum and maximum expected regrets for

the different strategies and select a product by applying the minimax criterion to the maximum expected regret of the different strategies. Compare your answer with that given to Question 22. Can any general conclusions be drawn from the comparison?

3. Discuss the circumstances in which the minimax criterion applied to the maximum expected regret of different strategies is particularly suitable for product selection.

FORECASTING

- Introduction
- Trend Analysis in Technological Forecasting
- Forecasting Jeans Sales in the UK using Decomposition Analysis
- Company Sales Forecasting using Exponential Smoothing
- Multiple Regression and the Market Size for New Cars

Assignments in Applied Statistics
Edited by S. Conrad
© 1989 John Wiley & Sons Ltd

Introduction

J. R. Sparkes* and S. F. Witt†

*Management Centre, University of Bradford, and †Department of
Management Science & Statistics, University College of Swansea

Business planning is concerned with making decisions that are based upon some assessment of the future. Forecasting is a means of helping business to cope with the uncertainty of the future. The more accurate the forecast, the more useful it will be. However, precision in forecasting is not always the most important thing. Accurate forecasts of sales, for example, are very important in the short run so far as production scheduling is concerned, but forecasting sales, say ten years ahead, does not need the same kind of precision. The more important feature in the long run is the trend and probable range of sales. The purpose of such forecasting is to put the business in a better position to anticipate the future. In this way forecasting, while not a means of avoiding uncertainty, serves to *lessen* uncertainty about the future.

Although it is difficult to generalize, companies' planning procedures are usually built around different planning horizons. The *short term*, for example, generally covers a period of one or two years ahead, the first year often being the annual budget against which management performance is appraised. *Longer term* planning requires a more strategic appraisal, seeking to identify those sectors where major opportunities or threats may lie several years ahead.

How short-term, medium-term and long-term plans fit into the planning cycle is clearly a matter for the individual organization, but there must invariably be overlap between the different planning horizons. Even for short-term planning a check with long-term trends is often useful. While we need to remember, therefore, that there are different planning horizons, we equally need to realize the links between them. The role of forecasting in business planning should be seen against this interdependence of the various planning horizons.

THE NEED FOR FORECASTING

Everybody makes forecasts, because forecasting derives from the economic problem that confronts every individual. The problem is that an individual's wants are unlimited while means are limited. This problem of scarcity, if only that of allocating time between competing alternative uses, gives rise to the need for choice, and this decision taking process is based on some assessment of likely results. Even if we do not make explicit forecasts, we at least make assumptions about the future in connection with our current actions.

Similarly the businessman has no choice but to forecast. The production manager needs sales forecasts in order to plan future production; the financial manager has to budget for cash needs; the personnel manager needs forecasts of the availability of labour for purposes of manpower planning, and so on. Whatever an organization's objectives, its efforts to achieve them inevitably call for the formulation of expectations about the future. All such forecasting is directed at reducing the uncertainty surrounding the future. However, in the context of business planning the need for some *explicit* forecasting procedure emerges. Among all the possible forces that can affect a business, those significant to the individual business need to be identified. Only by systematic analysis of the future can a company isolate those variables that are important to itself; and it can do this only if it first evaluates the accuracy of past estimates as a basis for correcting forecasting procedure. In this respect the organization needs not only to record what is happening at the present but also to have some means of judging what has happened in the past.

LIMITATIONS OF FORECASTING

Experience of the past is the best guide we have to the future. The adequacy and reliability of recorded data is therefore a particularly important problem, and one of the major limitations on business forecasting. However, its very existence alerts us to the importance of judgement in forecasting, especially in interpreting data, in selecting methods of analysis and in applying them to specific problems.

There is a very wide range of methods of forecasting, from the completely intuitive to the highly mathematical, but whatever the method employed all forecasters face three major problems:

1. *Identification.* For almost any important forecast the causal factors entering into the forecast are too numerous for them all to be taken into account in any systematic fashion. The identification problem is therefore one of isolating the key strategic variables causing change.
2. *Measurement.* Measurement is complicated by the fact that relations that have been observed to hold in the past cannot be assumed to hold precisely in the future.
3. *Sequence.* The sequence of cause and effect between relations is subject to variation with changes in the environment within which they operate.

Long-range planning requires that the organization continuously seeks to identify, first, those areas in which, as future changes take place, major opportunities or threats may lie and, second, those areas of its activity that are potential sources of strength or weakness. In which of its existing areas of activity does a company want to increase its commitment? From which does it want to break away? What *new* areas of activity are likely to offer opportunities? How do these fit in with existing activities? In seeking to answer such questions there is no simple formula to tell us which variables to forecast. It is for each organization to determine systematically in the light of its own strategic objectives which factors have a high probability of significant impact on the company. Similarly, the purpose for which a forecast is made will usually determine which of the many forecasting techniques available to the forecaster is most appropriate to use. In the assignments that follow, a selection of forecasting techniques has been chosen to illustrate quite different applications.

SELECTION OF FORECASTING METHOD

In general a forecasting method involves the following elements:

(a) choosing a suitable model;
(b) estimating the model;
(c) using the model for forecasting;
(d) assessing the forecasts produced;
(e) updating the model.

Choosing a suitable model

By model we mean the mathematical equations or procedures used to determine the forecasts. A very simple time series model, for instance, is the single exponential smoothing model. This says that

$$\text{Forecast for next period} = A \times \text{this period's actual value}$$
$$+ (1 - A) \times \text{forecast for this period} \quad (1)$$

where A is a constant. A rather more complex market model would be

$$\text{Sales} = A + B \times \left(\frac{\text{our price}}{\text{competitor's price}}\right)$$
$$+ C \times \left(\frac{\text{our advertising budget}}{\text{competitor's advertising budget}}\right) \quad (2)$$

where A, B and C are constants. This model is conceptually very different from the previous one since it ascribes causality. The ratio of our price to the competitor's and our advertising budget to the competitor's are assumed to determine sales. Models of this type are usually referred to as *causal* models.

Estimating the model

A forecasting model usually contains a number of constants, e.g. *A, B, C* in the examples above, whose values must be determined. The process by which their values are fixed is usually referred to as estimation. The most common approach to estimation is to select the parameter values in such a way that the forecasts produced are close to the corresponding actual values, which is achieved by selecting the parameter estimates so that the model fits *past* data well. It will be obvious, however, that there are many circumstances under which such an approach to estimation is inappropriate. It is unlikely to be useful in many strategic applications where we are interested in forecasting for an activity for which no equivalent has existed in the past.

Using the model for forecasting

Whereas choosing and estimating can claim substantial amounts of computer time, the production of forecasts once the forecast model has been obtained usually needs very few calculations. Nonetheless, it is not always easy to obtain the desired forecasts once the forecast model has been derived. The ease with which this can be done indicates how simple it is to use a particular computer forecasting package.

Assessing the forecasts produced

The most frequently used formal measures of forecast accuracy are mean square error, mean absolute error and bias. Mean square error is calculated simply as the average value of the square of the forecast error (the difference between the forecast and the actual value). Mean absolute error is the average of the numerical value of the forecast error. Bias is measured by the difference between the average of the actual value of the variable and the average of its forecast value, or it may be measured by the difference between the total of the actual values and the total of the forecast values. Usually it is considered desirable for the forecast to fluctuate on either side of the actual value. If this is the case, either measurement of bias will give a result close to zero.

What the manager needs is a reasonable estimate of how accurate a forecast method is likely to be in practice. This affects how we go about measuring forecast accuracy. In many types of short- and medium-term forecasting it is possible to assess how the forecasting model will work in practice by retaining the last few points of the data series and choosing and estimating the model using the remainder. The forecast model is then used to predict the values of the 'held-out' data points. Comparison of these predictions with the corresponding actuals provides a good measure of how the forecast method is likely to work in reality.

Updating the model

Choice and estimation of a forecast model will in general require both more computer time and a higher level of expertise than actually producing the forecasts from the model. For this reason, they are usually only carried out periodically, for example once every year for a monthly time series, rather than each time a forecast is to be prepared. In practice, forecast models sometimes perform better if they are updated for each period. The improvement in performance tends to be rather small, however, if the forecasting method is suitable for the application. Unless the losses due to inaccurate forecasts are large, updating the forecasting model annually is adequate.

Further determinants of forecasting method

We have already reviewed a number of factors that affect choice of forecasting methods. There remain, however, a number of more technical considerations that also help determine the choice. Many methods require more data than are readily available, as, for example, in the case of forecasting new product sales.

Besides accuracy, another factor to consider is the method's 'robustness'. Some methods are very sensitive to the assumptions made in applying them or to small changes to some of the data points. An example of this is S-shaped growth curves where small changes to the data in the early history of the time series can have a big effect on the values forecast. Obviously the results derived by such methods can be markedly affected by inexpert application. By contrast, other methods, for example some time series methods, are little affected by such maltreatment. In practice, it is hard to use a particular set of forecasts with any confidence, if one is aware that very different figures could be obtained with only slight changes to the way the forecasting method was applied. Robustness is accordingly of great practical concern.

It is often necessary to use a number of forecast methods together to achieve the forecast required. Thus it might be appropriate to forecast total industry sales of a product using a trend curve and then to forecast company sales using a regression model that used the industry sales forecast as one of its independent variables. Though there is a large number of forecasting packages available they are never completely comprehensive. Since there are often benefits in standardizing on the use of one package, for example being able to obtain advice on its use, compromising on a method that is somewhat less than ideal for a particular application may be justified.

Finally, there are two further related dimensions along which a forecasting method should be judged. First, there is the cost of applying it, including the computer time required for the choice and estimation of the forecast model. The costs of computation have fallen, but where there are many thousand forecasts to be prepared, as is often the case in consumer goods industries, banking and finance, etc., the use of complex methods requiring much computation may well be precluded. Obviously a similar situation exists with regard to the costs of data gathering, which nowadays are frequently much higher than the costs involved in actually computing the forecasts. This is currently the case, for example, in obtaining information from commercial computer databases.

Assignments in Applied Statistics
Edited by S. Conrad
© 1989 John Wiley & Sons Ltd

Trend Analysis in Technological Forecasting

J. R. Sparkes* and S. F. Witt†

*Management Centre, University of Bradford, and †Department of
Management Science & Statistics, University College of Swansea

OUTLINE

Trend analysis fits a particular type of trend curve to time series data and extrapolates to derive an estimate of the forecast variable. This assignment considers the rationale underlying trend analysis, the arguments for and against trend extrapolation, as well as some of the technical issues associated with trend analysis in the context of specific technological forecasting applications. A discussion of the limitations on the use of exponential growth curves in technological forecasting leads to consideration of two specific S-shaped curves widely used in fitting a mathematical function to historical data for the purposes of projecting the trend. These are the logistic curve and the Gompertz curve, the use of both in technological forecasting being considered. The latter part of the assignment extends the scope of technological forecasting using time series techniques from a narrow concern with the growth of functional capability to a broader interest in the adoption of technology as an important step towards integrating technological capability into business planning.

KEYWORDS

attributes, business planning, compound parameter, envelope curve, Fisher–Pry model, Gompertz curve, growth curves, logistic curve, parameters, precursors, S-curve, strategic forecasting, substitution curve, technological capability, technological forecasting, time series, trend analysis

The concept of a secular trend is useful in forecasting because many time series display long-term tendencies related to forces operating in the environment over long periods of time. Two familiar examples are population growth and technological development. The long-term tendency of a series of data may be represented by one of a number of different trend curves or patterns. The smoothness of the curve or pattern corresponds to the notion that 'trend' represents the continuous and gradual change caused by the workings of long-term factors.

Trend analysis fits a particular type of trend curve to a time series and then extrapolates to derive an estimate of its level at some future date. (See Granger, 1980, Ch. 2; Saunders, Sharp and Witt, 1987, Ch. 9.) In strategic forecasting, where trend analysis is widely used as an aid to management decision making about long-term plans, the trend level of sales, or whatever the application may be, will generally be more significant than a specific sales forecast. Thus, while a given mathematical method used to fit a particular type of trend curve to a time series produces one definite projection, the user of trend analysis must always bear in mind that, in what might otherwise appear a mechanical exercise, subjective judgement is usually necessary.

In this assignment, we consider some of the advantages and disadvantages of trend analysis in the context of a strategic forecasting area where its application has been widely and effectively used, namely technological forecasting. (See Granger, 1980, Ch. 10; Makridakis, Wheelwright and McGee, 1983, Ch. 13; Saunders, Sharp and Witt, 1987, Ch. 9.)

Technological change concerns long-term changes in the characteristics of products, processes and techniques. Technological forecasting is an attempt at predicting these future characteristics. Trend extrapolation may therefore seem of limited use in the context of a technological perspective where change is of the essence, because the trend almost certainly will bend. However, numerous studies have demonstrated how the developments of technological parameters, like non-technological ones, have followed smooth and orderly patterns. Furthermore, the sophistication of methods of trend extrapolation has increased considerably. Both of these considerations are illustrated in Figure 1.

Figure 1 demonstrates the smooth progression of technological advance in the context of the functional capability of the speed limits imposed by the computer's 'clock cycle'. At the beginning of the 1950s vacuum tubes used in computers required 10,000 nanoseconds (billionths of a second) to switch on or off. By the late 1950s, transistors cut switching time to one-hundredth of that. By the 1970s the silicon chip had again reduced switching time to one-hundredth of that of the transistor and in the 1980s the fastest circuits can switch on or off in less than one billionth of a second.

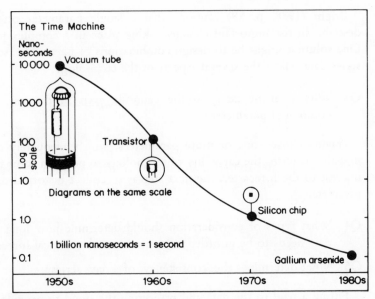

Figure 1
Developments in computer technology. (From *The Economist*, 3–9 April 1982, p. 129. *Reproduced by permission of* The Economist)

The *trend* in the increase in functional capability has remained steady despite the radical developments in computer technology since the 1950s. This provides confidence in being able to forecast that, whatever new technical approach is developed, it will achieve a certain level of functional capability by, say, the 1990s. Forecasting technology beyond the limits of the current state of the art is, on the basis of the trends shown, plausible.

In this assignment we consider the rationale underlying trend analysis, the arguments for and against trend extrapolation, and some of the technical issues associated with trend analysis in the context of specific technological forecasting applications.

The rationale for trend extrapolation in technological forecasting is that technical progress proceeds, as illustrated in Figure 1, in a fairly orderly manner, showing patterns of behaviour in past trends that will continue into the future.

Q1 In what circumstances do you consider it would be unreasonable to extrapolate a past trend when forecasting for technology?

The first stage of a trend projection is the choice of appropriate parameters which identify the requisite 'attributes' (characteristics) of interest to the forecaster.

Q2 What considerations should govern the selection of the parameters in a technological forecast?

Bright (1978, p. 88) cautions that a single parameter often fails to describe all the important changes taking place in a complex technology. One solution might be to design combinations of parameters so that time series data reflect the several aspects of the technology.

Q3 Why is it necessary to be cautious even in the use of such a compound parameter?

Having chosen one or more parameters, which adequately describe in measurable terms the capability of technology to perform some function of interest to the forecaster, the next step is to collect historical data for the parameter(s).

Q4 What types of consideration should determine how long a series of data needs to be in order to be used in technological forecasting?

Q5 What determines the 'consistency' of a data series?

Fitting a tend to the data and projecting the trend to provide a forecast requires that the forecaster establishes the trend line which best represents the past behaviour of the time series and extrapolates it in a way which is most appropriate for the future. Extrapolating a trend indefinitely into the future is fraught with danger. The further ahead one tries to project a trend, the less likely it is that factors which influenced past behaviour will remain unchanged. Thus trend analysis using time series data is at best a guide to the short- to medium-term future. (What is 'long-term' depends very much on the parameter being forecast.)

Q6 Suggest alternative techniques which, used on their own or in conjunction with trend analysis, might provide longer term technological forecasts.

Because the development of many technological, as well as non-technological, parameters has followed a smooth pattern, their progress has frequently been characterized as exponential.

Q7 Discuss whether the implied assumption in the use of an equation of the form $y = y_0 e^{kt}$ (where y is the value of the parameter, t is time and k a constant) that exponential growth will continue indefinitely into the future is reasonable in a simple time series extrapolation of technological progress.

Because of the limitations on the use of exponential growth curves in technological forecasting, the exponential time series has been considered

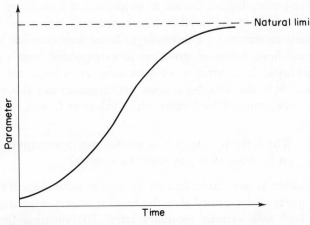

Figure 2
The S-curve

to represent only a part of a more generalized curve, the so-called S-curve shown in Figure 2. The S-curve is widely used in technological trend analysis, its shape reflecting the initial slow emergence of a new technology, followed by exponential growth to the inflection point when technological advance ceases to accelerate and then levels off.

Two specific S-shaped curves widely used in fitting a mathematical function to historical data for the purposes of projecting the trend are the logistic curve and the Gompertz curve. The logistic curve has its origins in biological growth processes observed by Raymond Pearl in the 1920s. The equation for the Pearl or logistic curve is

$$y = \frac{L}{1 + ae^{-bt}}$$

where L is an upper limit to the growth of the variable y and a and b are constants. ($y = 0$ at $t = -\infty$ and $y = L$ at $t = +\infty$.)

Q8 Use the second derivative of y with respect to time to find the values of y and t at the inflection point of the curve.

Q9 Draw a graph to depict the case where L, a and b are all equal to one, to show that the logistic curve is symmetrical about the inflection point.

Q10 Comment on the constants a and b in the light of this symmetry.

Q11 What will be the effect of a change in a?

When using logistic curves in technological forecasting, the upper limit to the growth of the variable, L, is sometimes determined by some known ultimate constraint on a technology. If the forecaster has knowledge of an ultimate limit, future progress can be extrapolated from a short time series of historical data, sufficient to determine values for a and b which provide a good fit to the data. Sometimes the forecaster can also use the historical data to determine L by finding which values of L, a and b give the best fit.

Q12 Why is the forecaster best advised not to attempt to predict the limit on the basis of a very short time series?

Another growth curve frequently used in technological forecasting is the Gompertz curve, named after the English actuary who originally proposed his 'law' as governing mortality rates. The equation for the Gompertz curve is

$$y = Le^{-be^{-kt}}$$

where L is an upper limit to the growth of the variable y, and b and k are constants.

Q13 Use the second derivative of y with respect to time to find the values of y and t at the inflection point of the curve.

Q14 Draw a graph to depict a Gompertz curve for which L, b and k are all equal to one.

Q15 Contrast both these answers with the answers previously obtained for the logistic curve.

There are numerous examples in the literature on technological forecasting to illustrate that data representing growth in the level of functional capability of a technology can be described by growth curves like the logistic and Gompertz curves. Apart from growth curves for functional capability, the technological forecaster can also make forecasts of the rate of adoption of a technology, based on historical data from the lower portion of the curve. Consider as an example the British market for video cassette recorders (VCRs). Table 1 shows actual and estimated sales of VCRs in Britain over the five years 1980–4. It also shows the percentage of British television households with VCRs.

When forecasting how the growth of a new technology will approach its upper limit, it is often simpler and more reliable to forecast the rate of adoption by households rather than unit sales. This is because the rate of adoption of a new technology will usually show the same general

Table 1
Britain's VCR growth (From *The Economist*, 5–11 February 1983, p. 35. *Reproduced by permission of* The Economist)

YEAR	VCRs (MILLION)	PERCENTAGE OF HOUSEHOLDS
1980	0.5	2.5
1981	1.5	7.5
1982	3.2	15.9
1983[a]	5.0	24.9
1984[a]	6.5–7.0	34.0

[a] Estimate.

behaviour as that represented by the S-curve. Adoption is never immediate, partly because of the economics of providing the new technology in the first place and partly because some potential adopters prefer to let others try something new first. The *absolute* upper limit must, of course, be 100 per cent of households, although the *effective* upper limit may be reached well before that level.

Q16 Use a Gompertz curve to produce a forecast for VCRs up to 1990. What upper limit to the level of adoption does your forecast suggest?

Extending the scope of technological forecasting using time series techniques from a narrow concern with the growth of functional capability to a broader interest in the adoption of technology is an important step towards integrating technological capability into business planning. Business growth opportunities depend upon such factors as the rate of adoption of a new technology. Business planning also requires some assessment of the rate at which a new technology will displace an existing one. This is of economic as well as technological significance and places the technological forecast in the context of the total forecast required in business planning. With this in mind, we can consider another application of the growth curve in technological forecasting. The 'substitution' curve forecasts the rate at which a new technology is substituted for an existing one.

Fisher and Pry (1971) developed a substitution model of technological change which is still widely used. Their hypothesis is that once a new technology begins to replace an existing one, it will proceed towards complete substitution. The substitution rate formula is

$$f = \tfrac{1}{2}[1 + \tan h \propto (t - t_0)]$$

or
$$\frac{f}{1 - f} = \exp 2 \propto (t - t_0)$$

where f = fraction substituted by new technology

$\propto$ = half the annual fractional growth in the early years

t_0 = time at which $f = \frac{1}{2}$

There are numerous examples of the application of the Fisher–Pry model to the substitution process. (See Bright, 1978, and Jones and Twiss, 1978, for examples and exercises using the methodology.)

Q17 Discuss the potential use of substitution techniques in the wider context of business planning and decision making.

Q18 What other types of trend extrapolation might most usefully be combined with substitution techniques?

Q19 Discuss the view that the Fisher–Pry model must be used with caution when forecasting the growth of a new technology.

Functional capability may progress through a succession of different technological advances. Just as the S-curve's progress within a particular technology enables time forecasts to be made for that technology, so it has

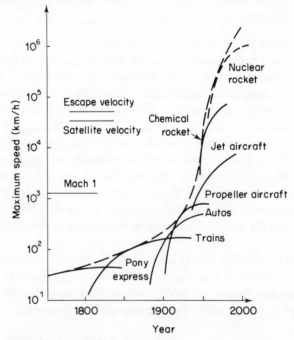

Figure 3
The envelope curve for speed. (From Jantsch, 1967. *Reproduced by permission of OECD*)

been suggested that a similar pattern of development may exist for the progress of an attribute through a series of new technologies. The outcome of such a relationship is a curve roughly tangential to the successive S-curves, the so-called 'envelope' curve, which is itself often S-shaped. This is well illustrated in Figure 3 for speed.

The methodology for projecting the envelope curve is essentially the same as for time series and growth curves. It is, however, likely to require historical data over a much longer time scale. Jantsch (1967, p. 163) suggests that in many practical cases the larger systems represented by envelope curves show more stable progress than their individual component technologies.

Q20   Suggest possible reasons for this alleged greater stability.

Perhaps the most important practical use of envelope curves is the possibility they offer of anticipating technological change. S-curves for individual technologies show the inability of a given technology to continue to support the progress exemplified in the envelope curve. The forecaster and decision maker will be interested in the displacement of an existing technology by a new one. Jantsch suggests that a deeper scrutiny of technologies in an early stage of development may lead to early recognition of the new 'growth' technology. Such scrutiny raises questions not only about the demise and advent of particular technologies but also about the resulting market impacts.

Q21   Explain how the example shown in Figure 4 illustrates the possible advantages as well as the possible disadvantages of using an envelope curve to predict the advent and, in broad terms, the effect of technological breakthrough.

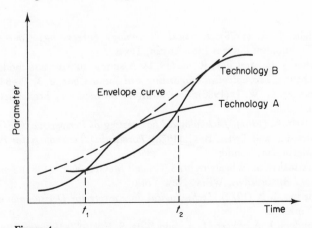

Figure 4
Technological breakthrough

Trend extrapolation to forecast technology by analysing 'precursors' uses the correlation of progress trends between two technologies, one of which acts as a leading indicator for the other. The most valid applications of this type of trend extrapolation are usually considered to be those for which advances in the precursor technology can be applied to the 'lagged' technology. A widely used example is that comparing the maximum speed of military combat aircraft and civilian transport aircraft. (See, for example, Jantsch, 1967, p. 159f; Martino, 1972, p. 150f.) One would reasonably expect advances made in combat aircraft technology to yield potential spinoff to transport aircraft, the former then acting as a precursor of developments in the civilian sector. Both authors show a steeper slope of the combat aircraft speed trend than of the transport aircraft speed trend, implying an increasing time lag between the two.

Q22   How might the observation that the time lag increases be explained?

Q23   Does the increasing time lag portend a breakdown in the precursor relationship?

Q24   Under what circumstances, if any, would it be reasonable to forecast the 'lagged' technology to levels not yet achieved by the precursor technology?

REFERENCES

Bright, J. R. (1978). *Practical Technology Forecasting: Concepts and Exercises*, Technology Futures Inc., Austin, Texas.

Fisher, J. C., and Pry, R. H. (1971). A simple substitution model of technological change, *Technological Forecasting and Social Change*, 3, 75–88.

Granger, C. W. J. (1980). *Forecasting in Business and Economics*, Academic Press, London.

Jantsch, E. (1967). *Technological Forecasting in Perspective*, OECD, Paris.

Jones, H., and Twiss, B. C. (1978). *Forecasting Technology for Planning Decisions*, Macmillan, London.

Makridakis, S., Wheelwright, S. C., and McGee, V. E. (1983). *Forecasting: Methods and Applications*, Wiley, New York.

Martino, J. P. (1972). *Technological Forecasting for Decisionmaking*, Elsevier, New York.

Saunders, J. A., Sharp, J. A., and Witt, S. F. (1987). *Practical Business Forecasting*, Gower, Aldershot.

SUPPLEMENTARY QUESTIONS

1. Critics of trend extrapolation might claim that there is no proof that past forces will continue to support a trend. Bright (1978) retorts that 'the burden of proof is on the critic'. How would you seek to justify such an assertion?
2. How would you determine for any given time series whether a Gompertz curve or a logistic curve will fit better?
3. Review briefly the process of technological innovation as a means of explaining the nature of technical progress reflected in S-curves.
4. A recent study examining 90 forecasts for new products, new markets and emerging technologies that were made public between 1960 and 1980 revealed that 53 per cent failed (Berenson and Schnaars, 1986). Compare the table of mistaken growth market forecasts with that of successful growth market forecasts and suggest why some of the forecasts were mistaken.

FURTHER REFERENCE

Berenson, C., and Schnaars, S. P. (1986). Growth market forecasting revisited: a look back at a look forward, *California Management Review*, 28(4), 71–88.

Assignments in Applied Statistics
Edited by S. Conrad
© 1989 John Wiley & Sons Ltd

Forecasting Jeans Sales in the UK using Decomposition Analysis

J. R. Sparkes* and S. F. Witt†

*Management Centre, University of Bradford, and †Department of
Management Science & Statistics, University College of Swansea

OUTLINE

A statistical series of observations over successive periods can yield several
categories of change. In this assignment, decomposition analysis is used to
break down a time series into the different types of movements such series
can produce. Such decomposition then provides a basis for forecasting.
The assignment identifies four basic categories of time series movements:
those attributable to the trend factor, the cyclical factor, the seasonal factor
and irregular or random changes. It examines statistical methods relating to
each of these components in the context of sales of pairs of jeans by UK
manufacturers and requires forecasts to be prepared. The supplementary
questions offer further applications of decomposition analysis and broaden
the discussion to the use of cyclical indicators in forecasting.

KEYWORDS

additive relationship, cyclical fluctuation, cyclical indicators,
decomposition, deseasonalized data, forecasting, irregular change,
least squares, logistic curve, method of simple averages, multiplicative
relationship, random change, ratio-to-moving-average method,
ratio-to-trend method, seasonal index, seasonal variation, secular
trend, specific seasonal, time series, typical seasonal

The observed values of a time series are usually the result of several influences. In this assignment we are concerned with the relative merits and demerits of the various methods that may be used in decomposing a time series. Isolating and measuring those parts of a time series that are attributable to each of the components of the series is the major purpose of decomposition analysis. (See Firth, 1977, Ch. 4; Makridakis, Wheelwright and McGee, 1983, Ch. 4.) Customarily time series variations are considered to be the result of four basic influences: secular trend, cyclical fluctuations, seasonal variations, and irregular or random changes.

The assignment will examine statistical methods relating to each of these components in the context of a particular example. Before that is done, however, we have to decide on the form of the decomposition model to use. Classical decomposition analysis usually assumes a multiplicative relationship between the four components of a time series. Thus if the trend component is denoted by T, the cyclical component by C, the seasonal component by S and the irregular or random component by I, we can express each observation, V, of a time series as

$$V_t = T_t \times C_t \times S_t \times I_t \quad \text{at time period } t$$

This multiplicative form of the model assumes that any particular value in a time series is the product of factors attributable to the various components.

Q1 Does the multiplicative approach have greater merit than an additive approach which sees each observation of a time series as being the sum of the four components?

Q2 Might the effects of the various components be neither entirely additive nor multiplicative but combined in any one of a large number of ways?

Q3 Some authors omit completely the cyclical component from the classical decomposition model, which is then presented in multiplicative form as $V_t = T_t \times S_t \times I_t$. What justification might there be for doing this?

A UK manufacturer of jeans, concerned about recent trends in the home market, is alarmed at the prospect of a moderate decline in the jeans business predicted by the chief financial officer of America's VF Corporation and reported in *The Economist*, 2–8 August, 1986. (See Figure 1.)

You will be required to compute the long-term trend of jeans sales by UK manufacturers, advise on seasonal factors and assess the importance of the cyclical factor in the UK market. Data for the (estimated) monthly sales of pairs of jeans by UK manufacturers is given in Table 1.

AMERICAN JEANS

THE BLUES

New York

The jeans business in America has faded and shrunk. Customers, especially women, have turned to softer and lighter casual clothes. In response, the big three American producers are re-shaping, diversifying and, now, buying each other.

VF Corporation, America's biggest publicly-held clothing company with sales last year of $1.5 billion, makes Lee jeans. For some $790m, it is buying Blue Bell, a clothing firm with sales of $1 billion a year which makes Wrangler and Rustler jeans, Jantzen swimming costumes and other apparel. It will pay Blue Bell around $380m in cash and VF stock, and also assume some $414m in long-term debt, incurred when Blue Bell went private in a $470m buyout in 1984. The deal will increase VF's total debt to about 50% of its capital, compared with 23% at the end of last year.

With the purchase of Blue Bell, VF's share of the $6 billion American jeans market will almost double to 23%, putting it ahead of the current market leader, Levi Strauss, which has 21%. VF reckons the addition of Wrangler, America's second most popular brand, to its Lee range, third in the market, will help it to attack Levi Strauss in its strongholds on the west coast and the Rockies. Blue Bell is already active in Levi Strauss territory; VF is not.

VF also wants to reach new customers through Blue Bell's distribution channels—discount houses like the K Mart chain into which VF does not sell. The recently slimmed Levi Strauss is unperturbed. It says it will keep selling jeans in more upmarket outlets like department stores and not take on VF in the discount stores. Since 1984, Levi Strauss has closed 39 (mostly jeans) plants and cut its work-force, by 9,800. This followed the decline in its net profits from $195m in 1983 to $41.4m in 1984—in its last full year as a public company before it went private in a buyout in August 1985.

Demand for jeans started to boom about 20 years ago and sales grew by 10–15% a year throughout the 1970s. Americans bought more than 600m pairs in 1981 but only about 460m pairs in 1985. Foreign markets have also slumped. Mr. Jerome Wiggins, chief financial officer at VF, predicts that the jeans business will continue to 'decline moderately'.

Figure 1
Jeans sales in America. (From *The Economist*, 2–8 August 1986, pp. 55–56. *Reproduced by permission of* The Economist)

Table 1
Sales of jeans by UK manufacturers (in thousands)

	1980	1981	1982	1983	1984	1985
January	1998	1924	1969	2149	2319	2137
February	1968	1959	2044	2200	2352	2130
March	1937	1889	2100	2294	2476	2154
April	1827	1819	2103	2146	2296	1831
May	2027	1824	2110	2241	2400	1899
June	2286	1979	2375	2369	3126	2117
July	2484	1919	2030	2251	2304	2266
August	2266	1845	1744	2126	2190	2176
September	2107	1801	1699	2000	2121	2089
October	1690	1799	1591	1759	2032	1817
November	1808	1952	1770	1947	2161	2162
December	1927	1956	1950	2135	2289	2267

SEASONAL FACTORS

It will be necessary to decompose the time series by first constructing a seasonal index. In the unlikely event that all variations in a time series were due entirely to seasonal influences, the *specific seasonal* for a given month is an index number whose base is the mean monthly variate. To determine the typical effect of any particular month in the time series we define the *typical seasonal* (more commonly called the seasonal index) for a month as the mean of all specific seasonals for that month.

Q4   If a time series is influenced not only by seasonal variations but by other variations, the mean monthly values are no longer constant. Why, in this event, would the possible solution of using a different base for each year of the time series in calculating the specific seasonals not be satisfactory?

To be of practical use, a seasonal index, which is a measure of seasonal variation, must describe typical rather than specific patterns of variation. To overcome the problem referred to above, alternative methods may be used to calculate the seasonal variation in a time series. The most common are:

(a) the method of simple averages;
(b) the ratio-to-trend method; and
(c) the ratio-to-moving-average method.

The method of simple averages

In this method, the purpose of calculating the mean value for each month of a time series is to eliminate the irregular and cyclical influences which distort the seasonal pattern.

Q5 Why is this unlikely to be achieved fully using this method?

Q6 Is the method of simple averages consistent with a multiplicative form of decomposition model?

The ratio-to-trend method

In this method each observation in a time series is expressed as a percentage of its corresponding trend value. The method is considered especially useful when a sudden change in trend occurs.

Q7 What are the method's main limitations for the purposes of constructing a seasonal index?

The ratio-to-moving-average method

For a monthly time series, the *specific* seasonal for a given month is an index number whose base is the moving average that is centred at that month. The ratio-to-twelve-month-moving-average method relies on a monthly time series being affected by seasonal variations which repeat regularly year after year.

Q8 Is the method still valid if seasonal variations alter from one year to the next?

To calculate typical seasonals (a seasonal index) from the specific seasonals by taking the mean of all specific seasonals for a month might give undue weight to some specific seasonal observations that are extreme in comparison with the rest.

Q9 How might this difficulty be avoided? Calculate the seasonal index for the time series of Table 1.

Seasonal variations distort the underlying trend and cyclical factors in a series of data. Our jeans manufacturer would get a distorted impression if the seasonal variations in the sales of jeans were misinterpreted as being the result of other factors. Monthly changes in sales may carry all kinds of implications for policies on materials orders, employment levels, stock holdings and so on.

Q10   Deseasonalize the sales figures for 1985 and demonstrate the significance of the deseasonalized data by comparing the figures for the months of June and October.

June is traditionally a peak month for jeans sales, October traditionally a low one. A fall in sales would therefore have been expected. How does the actual fall compare with what could have been expected in accordance with the typical seasonal pattern? Suggest an explanation for your answer.

THE LONG-TERM TREND

The trend factor (T) indicates the long-term general direction of a time series. Frequent changes in a secular trend are inconsistent with the idea that it represents the regular, steady movement of a time series in the long run.

Q11   How would you attempt to recognize whether a sudden change in the trend of a time series was real?

If a straight line trend is indicated by a time series, the most widely used procedure for fitting the data is by the method of least squares. Since the secular trend disregards seasonal influences, the calculation can be further simplified by condensing the data in such a way as to eliminate short-term variations.

Q12   Express the monthly data in Table 1 as a yearly time series and find the corresponding equation of the trend line.

The report from *The Economist* suggests that the trend of jeans sales may not be described adequately by a straight line.

Q13   Suggest and evaluate which of the many types of curves used in fitting trends might be the most useful in this instance.

The logistic curve has been widely used in the study of various types of growth. Originating in a zoological study, the logistic curve is based on an analysis of growth that revealed that initially the rate of growth is proportional to the size of population. Later, however, some factor begins to cause growth to decelerate—the so-called 'critical' point on the growth curve. After the critical point has been determined, the saturation level can be estimated.

Q14 Is a logistic curve helpful in explaining the trend of jeans sales in Table 1? (See, for example, Alder and Roessler, 1977, pp. 290–2.)

Q15 What conclusions do you draw from an attempt to fit the observations in Table 1 with a logistic curve?

THE CYCLICAL FACTOR

We have already indicated that some forecasters ignore the cyclical factor completely. This can be justified if the cylical component is so small as to be of little or no consequence. Sometimes, however, it is ignored because it is the most difficult component to identify. The difficulty of forecasting cyclical patterns in economic and business activity arises because, among other things, the causes of cyclical movements are generally not well understood, and the duration and amplitude of cycles vary from one oscillation to another. This makes projections of future occurrence especially hazardous. For this reason companies often resort to forecasting cyclical fluctuations by first forecasting or otherwise using conditions in their industry or the wider economy as indicators to company forecasting. Such indicators now play an important role in forecasting the UK economy's likely direction, as they have done for many years in the United States. (See Granger, 1980, Ch. 7; Sparkes, 1981; Makridakis, Wheelwright and McGee, 1983, Ch. 12.)

Q16 What criteria should determine the choice of series to be used as statistical business indicators?

Q17 What are the main problems associated with the use of indicators in forecasting?

Q18 Tabulate the cyclical factor and calculate the mean value for each year, disregarding the extreme values in each year. Comment on the movement of the average cyclical factor over the period of the data given. Would a subjective estimation of the future cyclical factor cause you to modify the computed value in any forecast you make?

Q19 Prepare forecasts of UK monthly sales for 1987. What qualifications would you add in presenting your forecast to the manufacturer?

Q20 Prepare a forecast of *annual* jeans sales for UK manufacturers in 1989. Does your forecast substantiate the predicted moderate decline in the jeans business for the United Kingdom?

REFERENCES

Alder, H. L., and Roessler, E. B. (1977). *Introduction to Probability and Statistics*, W. H. Freeman, San Francisco, California.

Firth, M. (1977). *Forecasting Methods in Business and Management*, Edward Arnold, London.

Granger, C. W. J. (1980). *Forecasting in Business and Economics*, Academic Press, London.

Makridakis, S., Wheelwright, S. C., and McGee, V. E. (1983). *Forecasting: Methods and Applications*, Wiley, New York.

Sparkes, J. R. (1981). The cyclical indicator approach to forecasting. *Managerial Finance*, 7(1), 10–15.

SUPPLEMENTARY QUESTIONS

1. Why is it considered necessary to use a long run of data in an analysis of secular trend? Illustrate your answer by using data from suitable published sources.

2. You are given the following photocopy agency sales data:

Photocopying agency sales (in thousands of £s)

	YEAR 1	YEAR 2	YEAR 3	YEAR 4	YEAR 5
January	18	21	25	35	42
February	18	25	29	39	37
March	18	25	29	27	25
April	18	21	33	26	29
May	20	16	24	16	
June	14	27	38	38	
July	32	41	44	68	
August	24	31	35	55	
September	15	19	34	38	
October	31	41	53	32	
November	26	28	49	50	
December	31	32	58	60	

Using decomposition methods, prepare forecasts for the remainder (May–December) of year 5.

3. Compile a database using *Business Monitor* monthly statistics (PM or SDM series). Analyse by decomposition the series you construct by calculating the seasonal factors, the trend and the cyclical factor, using the data up to and including the penultimate year of the series. Prepare forecasts for each month of the following year. How do these compare with the published data?

4. Examine the cyclical indicators for the UK economy available for use in an appropriate issue (or issues) of *Economic Trends*. (For an explanation of cyclical indicator methodology, see, in particular, CSO, 1975, 1976.) Do their timing characteristics suggest any usefulness in predicting turning points in jeans sales?

FURTHER REFERENCES

CSO (1975). Cyclical indicators for the United Kingdom economy, *Economic Trends*, **257**, 95–110.

CSO (1976). Changes to the cyclical indicator system, *Economic Trends*, **271**, 67–9.

Assignments in Applied Statistics
Edited by S. Conrad
© 1989 John Wiley & Sons Ltd

Company Sales Forecasting using Exponential Smoothing

J. R. Sparkes* and S. F. Witt†

*Management Centre, University of Bradford, and †Department of Management Science & Statistics, University College of Swansea

OUTLINE

Exponential smoothing models provide a relatively simple set of forecasting methods which tend to perform well in practice. They are cheap to compute and therefore are well suited to high volume applications. In this assignment several smoothing methods are considered; single exponential smoothing, Trigg and Leach's adaptive smoothing, Brown's double exponential smoothing, Holt–Winters' double exponential smoothing and Brown's triple exponential smoothing. The rationale underlying exponential smoothing and the applicability of the various smoothing methods to data series exhibiting differing characteristics are considered. Exponential smoothing models are examined in the context of forecasting company sales for several product lines. The supplementary questions include comparisons of the accuracy of forecasts generated by different exponential smoothing methods (after transformation of the original data series where appropriate).

KEYWORDS

adaptive smoothing, Brown's double exponential smoothing, Brown's triple exponential smoothing, changing variability, data transformation, exponential smoothing, first value, forecast error, forecasting performance, Holt–Winters' double exponential smoothing, robustness, seasonality, single exponential smoothing, stationary series, steps, time series, trend

Exponential smoothing models are examples of univariate time series models. They provide a relatively simple set of forecasting methods that tend to perform well in practice. As they are cheap to compute, smoothing models are well suited to high volume applications. Provided that the previous period's parameters are given as a starting point, smoothing models can be cheaply re-estimated each period, and this procedure tends to give better forecasts. The simplest smoothing method is single exponential smoothing; more complicated methods are adaptive smoothing, Brown's double exponential smoothing, Holt–Winters' double exponential smoothing and Brown's triple exponential smoothing. (See Makridakis and Wheelwright, 1978, Chs. 5–7 and 10; Thomopoulos, 1980, Chs. 4, 5 and 10; Makridakis, Wheelwright and McGee, 1983, Ch. 3; Saunders, Sharp and Witt, 1987, Chs. 2 and 3.)

This assignment examines the use of exponential smoothing models in the context of forecasting company sales. A firm manufactures various products and wishes to obtain forecasts of likely levels of future sales of these products.

SINGLE EXPONENTIAL SMOOTHING

The single exponential smoothing model in effect attempts to reduce forecast error by correcting last period's forecast by a proportion of last period's error:

$$F_{t+1} = F_t + k(x_t - F_t) \qquad (1)$$

where x_t = the value of a time series in period t

F_{t+1} = single exponential smoothing forecast of x_{t+1}
(i.e. the forecast for one period ahead)

k = constant, such that $0 < k < 1$

Equation (1) states that the forecast for period $t + 1$ is given by the forecast for period t plus a proportion (k) of the forecast error for period t. This equation may be rewritten to give

$$F_{t+1} = kx_t + (1 - k)F_t \qquad (2)$$

Q1 In order to fit the model to the data a forecast for the first value of the time series, F_1, is required. How may this be obtained? It is also necessary to choose a 'best' value for k. Discuss in detail how this may be derived.

Q2 Commonly used measures of overall forecasting performance are mean absolute error and mean square error. Comment on the advantages and disadvantages of these two measures.

Q3 Single exponential smoothing methods are only applicable to stationary series, that is to data without steps, trend or seasonality and with constant variance. If a trend were present in our time series, describe how we should transform the data so that we could subsequently apply the single exponential smoothing technique. If seasonality were present in our time series, discuss the necessary data transformations prior to forecasting using single exponential smoothing. Finally, how may a data series exhibiting changing variability be rendered stationary in the variance?

Table 1 gives data on monthly sales over a five-year period for product A. The company wishes to forecast monthly sales of the product in year 6.

Q4 Examine the data series given in Table 1 for the presence of non-stationarity in the variance. If the variability of the series appears to be changing with time, transform the series to render the variance constant.

Q5 Examine the data series derived in Question 4 for the presence of non-stationarity in the mean. If a trend is found to be present, transform the series to render it stationary.

Table 1
Sales of product A

	YEAR				
MONTH	1	2	3	4	5
January	116,879	95,492	114,663	161,330	98,106
February	98,278	115,282	107,617	101,819	111,575
March	94,296	91,322	207,346	135,435	107,711
April	144,179	81,978	120,273	97,811	85,363
May	95,916	75,954	117,167	111,706	70,887
June	91,973	86,867	123,525	208,515	106,390
July	54,148	51,908	147,767	52,369	56,692
August	70,848	97,810	179,166	161,028	67,102
September	86,000	82,093	152,006	87,258	53,415
October	126,354	109,856	116,321	150,300	124,616
November	108,093	120,602	159,432	93,827	57,106
December	72,775	133,438	91,848	120,555	79,260

Q6 Examine the data series derived in Question 5 for the presence of seasonality. If seasonal fluctuations are found to be present, transform the series to eliminate the seasonal component.

Q7 Use the data series derived in Question 6 to estimate a 'best' value for k in model (2). Generate monthly sales forecasts for product A in year 6 using the single exponential smoothing model (2).

Q8 Increase the value of k used in Question 7 by 0.05 and generate a new set of monthly sales forecasts for product A in year 6. Decrease the value of k used in Question 7 by 0.05 and calculate a new set of forecasts. Comment on the robustness of the model.

Q9 Omit the last six data points from the series derived in Question 6. Use the remainder of the time series to generate monthly sales forecasts for product A in year 6. Comment on the robustness of the model.

ADAPTIVE SMOOTHING

A modification of the single exponential smoothing model is adaptive smoothing. Here, instead of using a constant value for k, the value of this parameter changes in accordance with the forecast error. When the forecast error is large, the value of k in equation (2) is set close to 1 so that the forecast adjusts rapidly towards the previous actual value. When the forecast error is small, the value of k is set close to 0 so that the forecast remains almost unchanged. k is set equal to the absolute value of the ratio of the smoothed forecast error to the smoothed absolute forecast error. If the forecasts show consistent under- or over-forecasting the ratio will be near 1, but where there is no consistent bias the smoothed error and hence the ratio will be near 0. The Trigg and Leach (1967) approach to adaptive smoothing is embodied in the following set of equations:

$$E_t = x_t - F_t \tag{3}$$

$$SE_t = aE_t + (1 - a)SE_{t-1} \tag{4}$$

$$SAE_t = a\,|E_t| + (1 - a)SE_{t-1} \tag{5}$$

$$k_{t+1} = \left| \frac{SE_t}{SAE_t} \right| \tag{6}$$

$$F_{t+1} = k_{t+1}x_t + (1 - k_{t+1})F_t \tag{7}$$

Table 2
Sales of product B

YEAR	SALES
1	4111
2	4051
3	4064
4	4198
5	4097
6	4135
7	4211
8	4280
9	4260

where x_t = the value of a time series in period t

F_{t+1} = adaptive smoothing forecast of x_{t+1}

E_t = forecast error in period t

SE_t = smoothed forecast error in period t

SAE_t = smoothed absolute forecast error in period t

k_t = smoothing constant in period t

a = constant, such that $0 < a < 1$

Adaptive smoothing models, in common with single exponential smoothing models, are only applicable to data series that are stationary in the variance and do not exhibit trend or seasonality. However, whereas single exponential smoothing models are also only applicable to data without steps, adaptive smoothing models can accommodate sudden upward or downward steps in a series. Table 2 presents annual sales data for product B over a nine-year period.

Q10   Plot the sales figures against time. In what year does there appear to be a change in the level of the series?

Q11   Use the data series given in Table 2 to estimate the adaptive smoothing model depicted by equations (3) to (7). Generate annual sales forecasts for product B for years 10, 11 and 12.

DOUBLE EXPONENTIAL SMOOTHING—BROWN'S MODEL

Whereas single exponential smoothing and adaptive smoothing are not suitable for time series containing a trend, this is not the case with double exponential smoothing models. Brown's (1963) double exponential smoothing model produces forecasts containing both a constant level term and a linear trend term. The equations for Brown's double exponential smoothing model are as follows:

$$S_t^1 = kx_t + (1 - k)S_{t-1}^1 \tag{8}$$

$$S_t^{11} = kS_t^1 + (1 - k)S_{t-1}^{11} \tag{9}$$

$$T_t = \frac{k(S_t^1 - S_t^{11})}{1 - k} \tag{10}$$

$$C_t = 2S_t^1 - S_t^{11} \tag{11}$$

$$F_{t+h} = C_t + hT_t \tag{12}$$

where x_t = value of a time series in period t

S_t^1 = single exponential smoothing function value in period t

S_t^{11} = double exponential smoothing function value in period t

T_t = estimate of the trend term in period t

C_t = estimate of the constant level term in period t

F_{t+h} = Brown's double exponential smoothing forecast of x_{t+h}
(i.e. the forecast for h periods ahead)

k = constant, such that $0 < k < 1$

Our firm manufactures product C for which it has twenty years of annual sales data. The figures exhibit a strong upward trend and are shown in Table 3.

Q12 As with single exponential smoothing, the system needs to be initialized in order to fit the model to the data. Describe how initial estimates of the constant level and trend terms may be obtained.

Q13 Calculate the trend and constant level terms given by equations (10) and (11) respectively, using the data given in Table 3. Generate forecasts of annual sales of product C for years 21 to 25 using Brown's double exponential smoothing model.

In general, the forecasts produced by Brown's double exponential smoothing model tend to be better than applying single exponential smoothing to a series that has been transformed to render it stationary in the mean.

Table 3
Sales of product C

YEAR	SALES
1	910
2	980
3	1140
4	1243
5	1250
6	1370
7	1510
8	1600
9	1809
10	1767
11	1931
12	2090
13	2309
14	2445
15	2786
16	3210
17	3049
18	3178
19	3608
20	3456

Q14 Transform the data set shown in Table 3 to eliminate the trend from
the series. Apply the single exponential smoothing method to the
de-trended series and hence generate forecasts for years 21 to 25.

DOUBLE EXPONENTIAL SMOOTHING—HOLT–WINTERS' MODEL

Brown's double exponential smoothing model is not suitable for use with
seasonal data, but the model of Holt (1957) and Winters (1960) using
double exponential smoothing is specifically designed to be used with time
series exhibiting seasonality. The Holt–Winters' model produces forecasts containing a constant level term, a linear trend term and seasonal
factors, and is given by the following set of equations:

$$C_t = \frac{k_1 x_t}{S_{t-L}} + (1 - k_1)(C_{t-1} + T_{t-1}) \tag{13}$$

$$T_t = k_2(C_t - C_{t-1}) + (1 - k_2)T_{t-1} \tag{14}$$

$$S_t = \frac{k_3 x_t}{C_t} + (1 - k_3)S_{t-L} \tag{15}$$

$$F_{t+h} = (C_t + hT_t)S_{t-L+h} \tag{16}$$

where x_t = value of a time series in period t

C_t = estimate of the constant level term in period t

T_t = estimate of the trend term in period t

S_t = estimate of the seasonal factor in period t

F_{t+h} = Holt–Winters' double exponential smoothing forecast of x_{t+h}

L = length of the seasonal cycle

k_i = constant, such that $0 < k_i < 1$ ($i = 1, 2, 3$)

Q15 In order to fit the model to the data, initial estimates of the constant level, the trend and the seasonal factors are required. The Holt–Winters' model can function with only a few data points provided that *good* estimates of the seasonal factors can be made. Suggest how such estimates may be obtained.

Table 4 shows monthly sales data for product D manufactured by the firm. The series covers a 40 month period.

Q16 Use the Holt–Winters' double exponential smoothing model to generate monthly sales forecasts for months 41 to 50 for product D using the data given in Table 4.

Q17 Apply appropriate transformation(s) to the time series given in Table 4 so that Brown's double exponential smoothing model becomes applicable, and then use this method to generate forecasts for months 41 to 50.

Table 4
Sales of product D

	YEAR			
MONTH	1	2	3	4
January	18	21	25	35
February	18	25	29	39
March	18	25	29	27
April	18	21	33	26
May	20	16	24	
June	14	27	38	
July	32	41	44	
August	24	31	35	
September	15	19	34	
October	31	41	53	
November	26	28	49	
December	31	32	58	

Q18   Apply appropriate transformation(s) to the time series given in Table 4 so that single exponential smoothing becomes applicable, and then use this method to calculate forecasts for months 41 to 50.

REFERENCES

Brown, R. G. (1963). *Smoothing, Forecasting and Prediction,* Prentice-Hall, Englewood Cliffs, New Jersey.

Holt, C. C. (1957). Forecasting seasonal and trends by exponentially weighted moving averages, Carnegie Institute of Technology Research Paper, Pittsburgh, Pennsylvania.

Makridakis, S., and Wheelwright, S. C. (1978). *Interactive Forecasting,* Holden-Day, San Francisco, California.

Makridakis, S., Wheelwright, S. C., and McGee, V. E. (1983). *Forecasting: Methods and Applications,* Wiley, New York.

Saunders, J. A., Sharp, J. A., and Witt, S. F. (1987). *Practical Business Forecasting,* Gower, Aldershot.

Thomopoulos, N. T. (1980). *Applied Forecasting Methods,* Prentice-Hall, Englewood Cliffs, New Jersey.

Trigg, D. W., and Leach, A. G. (1967). Exponential smoothing with an adaptive response rate, *Operational Research Quarterly,* **18,** 53–9.

Winters, P. R. (1960). Forecasting sales by exponentially weighted moving averages, *Management Science,* **6,** 324–42.

SUPPLEMENTARY QUESTIONS

1. Smoothing models with a non-linear trend are also available. Brown's (1963) triple exponential smoothing model estimates a constant level term, linear trend term and non-linear (quadratic) trend term. Discuss this exponential smoothing model in detail. Apply Brown's triple smoothing model to the annual sales data on product C given in Table 3 in order to generate sales forecasts for the product for years 21 to 25.

Table 5
Sales of product C

YEAR	SALES
21	3582
22	3774
23	3965
24	4111
25	4051

Table 6
Sales of product D

MONTH	YEAR 4	5
January		42
February		37
March		
April		
May	16	
June	38	
July	68	
August	55	
September	38	
October	32	
November	50	
December	60	

2. The Holt–Winters' double exponential smoothing model is suitable for data exhibiting both trend and seasonality, and thus the conditions under which it can be applied are similar to those for the classical decomposition model. What are the relative advantages and disadvantages of these two forecasting methods?

3. Actual sales of product C for years 21 to 25 are shown in Table 5. Comment on the accuracy of forecasts produced by Brown's triple exponential smoothing model, Brown's double exponential smoothing model and the single exponential smoothing model applied to the de-trended series.

4. Actual sales of product D for months 41 to 50 are given in Table 6. Comment on the accuracy of forecasts produced by the Holt–Winters' double exponential smoothing model, Brown's double exponential smoothing model applied to the de-seasonalized series and the single exponential smoothing model applied to the de-trended and de-seasonalized series.

Assignments in Applied Statistics
Edited by S. Conrad
© 1989 John Wiley & Sons Ltd

Multiple Regression and the Market Size for New Cars

J. R. Sparkes* and S. F. Witt†

*Management Centre, University of Bradford, and †Department of
Management Science & Statistics, University College of Swansea

OUTLINE

Multiple regression is a causal forecasting method, where the forecast
variable is specifically related to a set of determining forces. This is in
contrast to the other three forecasting assignments—decomposition, exponential smoothing and trend analysis—in which past history on the
forecast variable is simply extrapolated and those forces that caused the
particular pattern for the time series are disregarded. In this assignment the
use of multiple regression analysis is examined in the context of forecasting
the size of the new-car market in the United Kingdom. The explanatory
variables include income, new-car prices, interest rates, hire-purchase
restrictiveness and motor fuel shortages. The appropriate forms of the
variables are examined and the interpretation of the empirical results is
considered in detail. Problems caused by the violation of the classical
assumptions are discussed and corrective action is covered in the supplementary questions.

KEYWORDS

autocorrelation, causal forecasting method, classical linear regression
model, coefficient estimates, coefficient signs, confidence interval,
distributed-lag model, dummy variables, Durbin–Watson statistic,
explanatory variables, functional form, goodness-of-fit, multicollinearity,
multiple regression, new-car market, ordinary least squares, partial
adjustment process, statistical significance, time series, variable
correlation matrix

Non-causal forecasting methods simply extrapolate past history on the forecast variable and disregard those forces that caused the particular pattern for the time series. A major problem with forecasting by extrapolation is that it pre-supposes that the factors which were the main cause of growth in the past will continue to be the main cause in the future, which may be incorrect, and if this is the case the use of this technique will result in poor forecasts. Multiple regression is, by contrast, a *causal* forecasting method; here the forecast variable is specifically related to the determining forces. Such an approach to forecasting is also the only one which is of any value to a company (or government) if it wishes to explore the consequences of alternative future policies. (For a generally non-quantitative introduction to multiple regression and econometrics, see Stewart, 1976; Cassidy, 1981, Chs. 1–4.)

This assignment examines the use of multiple regression analysis in the context of the new-car market in the United Kingdom. (For further discussion see Witt and Johnson, 1986.) Reliable forecasts of sales of new cars are crucial for efficient planning by car manufacturers, component suppliers, distributors, customer-finance companies and other industries associated with the new-car market. Button, Pearman and Fowkes (1982, p. 8) point out that for these companies:

> The standard technique is for each firm, using a variety of econometric and statistical procedures, to forecast the future national demand for new cars . . . and then to determine their own share of the market.

This approach to forecasting the sales of a product by an individual firm thus involves the construction of a two-stage forecasting model. The first stage is a forecast of the total market demand for the product, that is a market size forecast. The second stage is a market share forecast—what share of the total market for the product the individual firm is likely to obtain. The company sales forecast is then given by

$$\begin{array}{c} \text{Company sales} \\ \text{forecast} \end{array} = \begin{array}{c} \text{market size} \\ \text{forecast} \end{array} \times \begin{array}{c} \text{market share} \\ \text{forecast} \end{array}$$

The assignment concentrates on forecasting the size of the new-car market in the United Kingdom.

Regression is used to estimate the quantitative relationship between the variable to be forecast and those variables which appear likely to influence the forecast variable. The estimation is carried out using historic data, and future values of the forecast variable are obtained by using forecasts of the influencing variables in conjunction with the estimated relationship. The process of forecasting by regression may be summarized as:

(a) Select those variables that are expected to influence the forecast variable and specify the relationship in mathematical form.
(b) Assemble data relevant to the model.
(c) Use the data to estimate the quantitative effects of the influencing variables on the forecast variable in the past.
(d) Carry out statistical tests on the estimated model to see if it is sufficiently realistic.
(e) If the tests show that the model is satisfactory then it can be used for forecasting.

The first stage in the regression process involves the selection of those variables that are expected to influence the forecast variable. In the case of a product which is sold to the final consumer, the standard market demand function includes the following explanatory variables:

$$Y = f(X_1, X_2, X_3, X_4, X_5, \ldots) \tag{1}$$

where Y = market demand for the product

X_1 = consumers' disposable income

X_2 = size of the population

X_3 = price of the product

X_4, X_5 = prices of substitute and complementary products

f = some function

The set of determining variables considered in equation (1) is by no means exhaustive. For example, the demand for goods that are frequently purchased on credit (such as cars) may be influenced by interest rates.

Q1 Many functional forms are possible for equation (1), but the two most common forms used in regression analysis are linear and log-linear (multiplicative). Why might it be desirable to specify a demand function in log-linear rather than linear form?

Q2 If equation (1) is a multiplicative relationship, comment on the implications of modifying the forecast variable to become market demand per capita (i.e. divided by population) and excluding population size from the set of explanatory variables. What are the advantages and drawbacks of such a modification in the context of regression analysis?

The specification of the multiple regression model for forecasting new-car sales is now considered in detail. The dependent variable is the number of new-car registrations expressed in per capita terms. The explanatory variables comprise real personal disposable income per capita, an index of new-car prices expressed in real terms, an interest-rate variable that reflects the cost of borrowing to finance car purchases and a hire-purchase restrictiveness variable relating to the minimum percentage deposit on new-car purchases. As the data to be used in estimating the model cover the period 1961 to 1981, an additional variable is included to represent the impact on new-car sales of the shortages of motor fuel which occurred as a result of the 1973 oil crisis.

Q3   What is the appropriate population deflator for the dependent variable? Comment on whether the total UK population should be used or a particular subset.

Q4   Why is it necessary to specify the income and price variables in real (constant price) terms?

The model is specified in log-linear form as follows:

$$\ln N_t = \alpha_0 + \alpha_1 \ln Y_t + \alpha_2 \ln P_t + \alpha_3 \ln H_t + \alpha_4 \ln i_t + \alpha_5 D_t + U_t$$

$$\text{for } t = 1, 2, \ldots, 21 \ (1 = 1961, \ldots, 21 = 1981) \quad (2)$$

where N_t = number of new-car registrations per capita in year t

Y_t = real personal disposable income per capita in year t

P_t = real price of new cars in year t

H_t = minimum percentage deposit on new cars in year t

i_t = rate of interest in year t

D = dummy variable representing the effects of the 1973 oil crisis

$$D_t = \begin{cases} 1 \text{ if } t = 14 \text{ or } 15 \text{ (1974 or 1975)} \\ 0 \text{ otherwise} \end{cases}$$

U = random disturbance term

$\alpha_0, \alpha_1, \ldots, \alpha_5$ = parameters to be estimated

Q5   Estimate the multiple regression model (2) by ordinary least squares using the data given in Table 1.

Q6   Interpret each of the coefficient estimates.

Table 1
Values of variables required for estimating multiple regression model (2)
(index form)

YEAR	ln N	ln Y	ln P	ln H	ln i
1960	3.05993	6.56498	4.95179	2.85762	1.67654
1961	2.97047	6.59250	4.92953	3.06058	1.73484
1962	3.01955	6.58659	4.85215	3.14716	1.58063
1963	3.27094	6.62260	4.69639	2.99573	1.38704
1964	3.42285	6.65454	4.66354	2.99573	1.62077
1965	3.35832	6.67267	4.62765	3.13070	1.85926
1966	3.30160	6.69159	4.60732	3.45758	1.86687
1967	3.34244	6.70441	4.61810	3.51453	1.81775
1968	3.34205	6.71877	4.64243	3.53893	2.01036
1969	3.21707	6.73030	4.64309	3.68888	2.05975
1970	3.32122	6.76261	4.64024	3.68888	1.97810
1971	3.49763	6.76250	4.61251	3.44713	1.77749
1972	3.73381	6.83819	4.57700	3.04452	1.74624
1973	3.72217	6.89811	4.54865	3.06666	2.28920
1974	3.43753	6.89497	4.57571	3.50646	2.51163
1975	3.38368	6.88969	4.60517	3.50646	2.34670
1976	3.45606	6.87482	4.60647	3.50646	2.40902
1977	3.47293	6.85372	4.67582	3.50646	2.18605
1978	3.65984	6.93770	4.73444	3.50646	2.20849
1979	3.72176	6.99878	4.75686	3.50646	2.61747
1980	3.59586	6.99402	4.72825	3.50646	2.79104
1981	3.56702	6.95731	4.69774	3.50646	2.58385

Sources: CSO, *Annual Abstract of Statistics*, HMSO, various issues. BSO, *British Business*, HMSO, various issues. SMMT, *Motor Industry of Great Britain* 1982.

Q7   Comment on whether the signs of the estimated coefficients are as expected.

Q8   Do the magnitudes of the estimated coefficients appear reasonable? Can you deduce from these magnitudes whether new cars are necessities or luxuries?

Q9   Comment on the statistical significance of each of the coefficients.

Q10 Obtain a 95 per cent confidence interval for α_1. Does this additional information assist in deciding whether new cars are necessities or luxuries?

Q11 Comment on the goodness-of-fit of the estimated model.

Q12 How may the model specification be improved?

The classical linear (or transformed linear) regression model incorporates various assumptions, and if these assumptions hold for the model under consideration, then the ordinary least squares estimators possess certain desirable properties. If, however, the assumptions are violated, the properties do not hold. One of the classical assumptions is that the various values of the error term are drawn independently of each other. If this assumption is not satisfied, that is the value of the error term in any particular period is correlated with the preceding value(s), this gives rise to the problem of autocorrelation of the disturbance term. (See Wonnacott and Wonnacott, 1979, Ch. 6; Johnston, 1984, Ch. 8; Saunders, Sharp and Witt, 1987, Ch. 6.)

Q13   Suggest reasons why the values of the error term U in equation (2) may not be drawn independently of each other.

Q14   What would be the consequences if autocorrelated disturbances were present in model (2)?

Q15   What is the value of the Durbin–Watson statistic for the estimated model? What does this value imply about the possibility of autocorrelation being present in the model?

A further classical assumption is that the explanatory variables are linearly independent of each other. If this condition is not satisfied, the problem of multicollinearity arises. Although the existence of an exact linear relationship is rare, it is quite common for there to be an approximate linear relationship among the explanatory variables in a multiple regression model. (See Wonnacott and Wonnacott, 1979, Ch. 3; Johnston, 1984, Ch. 6; Saunders, Sharp and Witt, 1987, Ch. 6.)

Q16   What would be the consequences if multicollinearity were present among the explanatory variables in model (2)?

Q17   What is the variable correlation matrix for the explanatory variables used in estimating model (2)? What conclusions can be drawn from this matrix regarding the presence of multicollinearity in the model?

Q18   Why may it be necessary to carry out further tests for multicollinearity?

The dependent variable in a demand function represents the *desired* demand for a good or service. Where it is assumed that the desired level of demand is achieved, then desired demand may be equated with actual sales [as in equation (2)]. In the case of the new-car market, however, this assumption may be unrealistic. Most car owners tend to receive income increases every year, but few replace their car with a new one every year, so a partial adjustment process may be postulated. (See Wonnacott and

Wonnacott, 1979, Ch. 6; Johnston, 1984, Ch. 9; Saunders, Sharp and Witt 1987, Ch. 6.) It is assumed that each increase in income results in an increase in the desired number of new cars, but within a given year consumers only move part of the way to the new desired level and do not actually reach it. The specification of the partial adjustment process is given by the following distributed-lag model:

$$\ln N_t - \ln N_{t-1} = \lambda(\ln N_t^* - \ln N_{t-1}), \qquad 0 < \lambda < 1 \qquad (3)$$

where N_t = actual number of new-car registrations per capita in year t

N_t^* = desired number of new-car registrations per capita in year t

λ = coefficient representing the speed of adjustment of the actual to the desired value

Q19 What would the adjustment process (3) imply if λ were equal to zero? What would it imply if λ were equal to unity?

Q20 How may equations (2) and (3) be combined to yield a single model in which the dependent variable, $\ln N_t$, is given as a function of a set of explanatory variables? In what respect does this multiple regression model differ from equation (2)?

Q21 Estimate the new regression model over the period 1961–81 using the data given in Table 1.

Q22 Interpret each of the estimated coefficients. Why may these be thought of as relating to the short term? How may one derive the impacts of the explanatory variables on the dependent variable in the long term (after complete adjustment of the actual level of new-car sales per capita to the desired level has taken place)?

Q23 How have the empirical results improved/worsened by incorporating a partial adjustment process in the new-car demand model? Discuss in terms of the estimated signs and magnitudes and the statistical significance of the coefficients, and the coefficient of determination.

Q24 Why may the Durbin–Watson statistic no longer be a reliable indicator of the presence of autocorrelation?

Forecasts of the dependent variable, $\ln N_t$, are generated by multiplying future values of the explanatory variables by their respective estimated parameters and then summing. This yields a point estimate of the forecasted dependent variable—the expected value of the forecast variable. In order to generate forecasts of new-car registrations it is therefore necessary to obtain forecasts of population size, income, new-car prices, the minimum percentage deposit on new cars and interest rates.

Q25   By what percentage are new-car sales in the United Kingdom likely
to increase over the period 1981–8 on the assumptions that over the
same period:

(a) population size decreases by 2 per cent,
(b) real personal disposable income per capita increases by 10 per
cent,
(c) the real price of new cars decreases by 5 per cent (owing to
improved productivity and greater competitive pressures),
(d) the minimum percentage deposit on new cars remains
unchanged,
(e) interest rates rise by one-fifth?

How is the forecast affected if income increases by 15 per cent rather
than 10 per cent over the period?

REFERENCES

Button, K. J., Pearman, A. D., and Fowkes, A. S. (1982). *Car Ownership
Modelling and Forecasting*, Gower, Aldershot.
Cassidy, H. J. (1981). *Using Econometrics*, Reston Publishing, Reston, Virginia.
Johnston, J. (1984). *Econometric Methods*, McGraw-Hill, New York.
Saunders, J. A., Sharp, J. A., and Witt, S. F. (1987). *Practical Business Forecasting*,
Gower, Aldershot.
Stewart, J. (1976). *Understanding Econometrics*, Hutchinson, London.
Witt, S. F., and Johnson, S. R. (1986). An econometric model of new-car demand
in the U.K., *Managerial and Decision Economics*, 7, 19–23.
Wonnacott, R. J., and Wonnacott, T. H. (1979). *Econometrics*, Wiley, New York.

SUPPLEMENTARY QUESTIONS

1. What are the assumptions underlying the classical linear regression
model?
2. Discuss the properties of the ordinary least squares estimators in the
context of the classical linear regression model.
3. Discuss appropriate corrective action to be taken if autocorrelation is
found to be present in a regression model.
4. Discuss possible corrective action to be taken if multicollinearity is
found to be present in a regression model.

5. The usefulness of the linear regression model can often be extended by the inclusion of dummy variables as explanatory variables in the model. These are qualitative variables which generally take the value 0 or 1 and may be used to represent factors which include the following:

   (a) *Temporal effects.* A behavioural relationship may shift between one period and another. For example, if we are using quarterly data to estimate a demand function, a seasonal pattern may be expected.

   (b) *Spatial effects.* A behavioural relationship may shift between one area and another. For example, a market demand function estimated jointly for UK and West German data may well generate a higher income elasticity for the United Kingdom (where the good is likely to be regarded as more of a luxury, i.e. a higher income elasticity is expected) than for Germany (where the good is likely to be regarded as more of a necessity, i.e. a lower income elasticity is expected).

   (c) *Qualitative factors.* The non-quantifiable characteristics of a good may be important determinants of consumer demand.

   (d) *Broad bands of quantitative variables.* Although numerical values of a variable are available, only broad groupings may be required. For example, the demand for sports cars may not be considered to depend on a person's precise age, but rather merely whether he or she is aged over 35 or less than or equal to 35.

   Illustrate the use of dummy variables and discuss the interpretation of parameter estimates.

6. Construct a model to explain the demand for beer. Assemble relevant recent data and estimate the model. Forecast annual beer demand for the next three years on the basis of alternative assumptions regarding movements in the explanatory variables.

7. The following market demand model for tobacco goods is specified:

$$\ln Y_t = \alpha_0 + \alpha_1 \ln X_{1t} + \alpha_2 \ln X_{2t} + \alpha_3 \text{HSA} + \alpha_4 \text{HSB} + \alpha_5 \text{HSC} + U_t$$
$$\text{for } t = 1, 2, \ldots, 21 \quad (4)$$

where Y_t = consumption of tobacco goods in year t

   X_{1t} = real personal disposable income per capita in year t

   X_{2t} = real price of tobacco goods in year t

   HSA, HSB, HSC = health scare dummy variables

   U = random disturbance term

   $\alpha_0, \alpha_1, \ldots, \alpha_5$ = parameters

During the period of the analysis three major health scares related to tobacco smoking occurred, and inspection of the data suggests that the

Table 2
Values of variables required for estimating multiple regression model (4)

YEAR	Y (MILLION lb)	X_1 (£)	X_2 (INDEX)
1	246.5	358.3	0.94
2	249.5	396.5	0.93
3	256.0	429.1	0.95
4	260.7	449.4	0.97
5	266.1	475.7	0.97
6	274.6	520.0	1.00
7	277.7	559.2	1.01
8	266.4	593.3	1.02
9	273.2	632.2	1.00
10	266.5	694.6	1.02
11	254.8	709.1	1.09
12	257.0	721.4	1.07
13	255.4	729.8	1.05
14	253.4	741.7	1.04
15	249.3	745.9	1.06
16	247.4	771.3	1.00
17	236.8	786.7	0.93
18	249.1	843.9	0.88
19	262.3	891.2	0.81
20	258.5	906.2	0.82
21	245.7	896.0	0.84

Sources: Tobacco Research Council, Research Paper 1, 7th edition. CSO, *Annual Abstract of Statistics*, HMSO, various issues.

reduction in demand as a result of each health scare was restricted to the year of the scare and the subsequent year. The health scare dummy variables therefore take the following forms:

$$HSA = \begin{cases} 1 & \text{for years 8 and 9} \\ 0 & \text{otherwise} \end{cases}$$

$$HSB = \begin{cases} 1 & \text{for years 10 and 11} \\ 0 & \text{otherwise} \end{cases}$$

$$HSC = \begin{cases} 1 & \text{for years 17 and 18} \\ 0 & \text{otherwise} \end{cases}$$

Estimate the multiple regression model (4) by ordinary least squares using the data given in Table 2. Interpret each of the coefficient estimates and comment on the signs and magnitudes of the estimated coefficients, paying particular attention to the dummy variable coefficients. Carry out appropriate statistical tests on the model and discuss the results.

MULTIVARIATE ANALYSIS

- Introduction
- Principal Components Analysis of Medical Records
- The Classification of Depression amongst Women
- Discriminant Analysis for Psychiatric Screening
- Understanding Pain through Multidimensional Scaling

Assignments in Applied Statistics
Edited by S. Conrad
© 1989 John Wiley & Sons Ltd

Introduction

B. S. Everitt

Biometrics Unit, Institute of Psychiatry, University of London

The data collected by today's researchers in all disciplines are almost always *multivariate*. This simply implies that more than a single measurement or observation has been made on each of the objects, subjects or patients under investigation. An example of a small set of multivariate data is shown in Table 1.

Such data are inherently more complex than the single variable, *univariate*, counterpart, simply because of the increased number of variables. Additional complexity is also introduced because of the possible relationships between the variables.

Methods for the analysis of multivariate data may be loosely grouped into *explanatory* or *confirmatory* techniques. Members of the former class 'explore' data in an attempt to recognize any non-random pattern or structure requiring explanation. At this stage, finding the question is often more important than finding the subsequent answer, the aim of this part of the analysis being to generate possibly interesting hypotheses for later study. Here formal methods designed to yield specific answers to rigidly defined questions are not required. Instead methods are sought that allow possibly unanticipated patterns in the data to be detected, opening up a wide range of alternative explanations. Such techniques are generally characterized by their emphasis on the importance of visual displays and graphical representations, and by the lack of any associated stochastic model, so that questions of the statistical significance of results hardly ever arise. The assignments in this section primarily concern this type of technique.

A confirmatory analysis becomes possible once a research worker has some well-defined hypothesis. Questions regarding the statistical significance of results now become perhaps of greater importance. For multivariate data, techniques such as Hotelling's T^2, multivariate analysis of variance, etc., fall into this category; such techniques are directly analogous to more familiar univariate methods such as Student's t-test and analysis of variance. Whilst these methods are of considerable importance in particular situations, they are not considered here; a useful

Table 1
Example of a set of multivariate data

INDIVIDUAL	HEIGHT (IN)	WEIGHT (LB)	EYE COLOUR	SMOKER/ NON-SMOKER
1	66	120	Blue	S
2	72	130	Green	S
3	70	150	Blue	NS
4	73	180	Grey	S
5	65	115	Brown	NS

non-mathematical description of such techniques is given in Hand and Taylor (1987).

PRINCIPAL COMPONENTS ANALYSIS

One of the first tasks with a large set of multivariate data is to summarize it in some way which will enable the researcher to grasp the most important relationships between variables. A useful first step is to seek some way to reduce the dimensionality of the data, and the method commonly employed is *principal components analysis*. The basic idea of the method is to describe the variation of the n objects or individuals in the original p-dimensional space in terms of a new set of uncorrelated variables which are *linear combinations* of the original variables. The new variables are derived in decreasing order of importance so that, for example, the first principal component accounts for as much as possible of the variation in the original data. The usual objective of this type of analysis is to assess whether the first few components account for most of the variation in the original data. If they do, then it is argued that they can be used to summarize the data with little loss of information, thus providing a simplification in the data by a reduction of its dimensionality. As well as providing a parsimonious description, principal component scores may prove invaluable in simplifying further analyses.

The assignment involving principal components deals with the analysis of a table of medical records. The aim of the analysis of these records is to explore similarities or differences between the doctors providing the records. The doctors appear to vary widely in the way in which they classify (diagnose) patients suffering from psychological distress and the aim of a principal components analysis of these data is to look at patterns of variation in their diagnostic practices. How do the doctors differ? Is there any sign of clustering of doctors or is there any evidence of any unusual or outlying diagnostic practices, and so on? Having carried out an initial analysis using principal components one then might wish to consider other methods of multivariate data exploration to supplement or even take the place of the principal components analysis. It is quite common, for example, to use methods of cluster analysis to supplement and to validate conclusions reached through the use of principal components analyses or analyses using other multidimensional scaling techniques. Alternative approaches might involve the use of factor analytic models and, particularly in the context of the present data set, the use of correspondence analysis. The student is encouraged to consider principal components analysis as just one example of several different approaches to multidimensional scaling.

CLUSTER ANALYSIS AND DISCRIMINANT FUNCTION ANALYSIS

An important component of virtually all scientific research is the classification of the phenomena being studied. The investigator is usually interested in finding a classification in which the items of interest are placed into a small number of homogeneous groups or clusters, within which items are relatively similar. At the very least the classification may provide a convenient summary of the multivariate data, but it will often yield much more than this. It will be an aid to memory and the understanding of the data and will facilitate communication between different groups of research workers. Often it will have important theoretical or practical implications: for example, the classification of chemical elements in the periodic table produced by Mendeleyev in the 1860s has had a profound influence on the understanding of the structure of the atom.

The classification process essentially consists of two separate but related steps. The first is the construction of a sensible and informative classification of initially unclassified set of objects. The second involves the derivation of rules for allocating objects to one of a number of previously defined categories. The first stage can be illustrated by the example of the field worker in archaeology who finds large numbers of objects such as stone tools, funeral objects, pieces of pottery, ceremonial statues and skulls: the worker would like to produce a classification of these objects since this might aid in discovering whether they arose from a number of different civilizations. If each object was described by a set of measurements, then one of the techniques of *cluster analysis* could be used to produce the required classification.

The cluster analysis assignment in this section involves a set of data collected by a psychiatrist during an investigation of depression amongst women. Diseases of the mind are more elusive than diseases of the body, and the classification of such diseases is in an uncertain state. A psychiatrist may collect a large amount of information on a sample of mentally ill patients and use this to try to determine whether a classification of the patients can be produced which has implications for diagnosis and treatment. There are often particular problems with psychiatric data and a number of these are encountered with the depression in women data.

The second stage of the classification process, rules for allocating individuals or objects to *a priori* defined classes can be illustrated by the example of a disease that can be diagnosed without error only by means of a postmortem examination. The physician would clearly like to develop a rule for allocating suspected cases to either the disease class or the no-disease class while the patient is still alive, so that, if necessary, appropriate action could be taken. Techniques applicable to this second stage are known as *discriminant function techniques*.

The data presented here to illustrate discriminant analysis are responses taken from samples of three classes of people: normal people, people diagnosed by psychiatrists as mildly ill and people diagnosed as severely ill. Each subject has given answers to 25 questions, and the aim is to formulate a classification rule, using these data, by means of which future subjects may be classified. Such an instrument would be useful for screening purposes.

Having formulated such a classification rule one needs to know how well it is likely to do: will it classify most future cases correctly or not? This is answered by estimating future misclassification rates.

Of course, real data often present problems not encountered in standard textbooks: missing values, peculiar distributions, contravened assumptions, etc. The data here are no exception and the reader's attention is drawn to some of these problems.

MULTIDIMENSIONAL SCALING

A frequently encountered type of data, particularly in the behavioural sciences, is the *proximity matrix* arising either directly, from experiments in which subjects are asked to assess the similarity of two stimuli, or indirectly, as a measure of the correlation or covariance of two stimuli derived from their raw profile data. Multidimensional scaling techniques attempt to uncover the pattern or structure of the stimuli as implied by their observed proximities (either similarities or dissimilarities) by representing such patterns as a simple geometrical model or diagram. To do this a set of coordinates is derived for each of the stimuli. Note that this technique differs from those described previously where data points were directly observable as n points in p-dimensional space; here only a function of the data points, their proximities, is observed. A simple example illustrating the objective of these methods would be to construct a map of a country given the road distances between each of a number of cities or towns in the country.

The multidimensional scaling assignment involves data collected during the development of an instrument for measuring pain, an important quantity in medical diagnosis. Subjects were asked to assess the similarity of adjectives describing various types of pain, and a similarity matrix was constructed for a set of 30 adjectives. By applying multidimensional scaling techniques the dimensions underlying judgement of pain can be investigated.

SUMMARY

Techniques for the analysis of multivariate data can be loosely grouped as *exploratory* or *confirmatory*. Members of the former class seek out interesting patterns in the data which may be suggestive of structure to be followed up in later investigations. Confirmatory techniques, on the other hand, are used to test specific hypotheses. The assignments in this section are primarily concerned with exploratory analysis using techniques such as cluster analysis and principal components analysis. The data sets for each of the assignments present problems such as missing values, peculiar distributions, etc., problems almost always encountered with real data. By being asked to analyse such data the student will hopefully gain greater insight into both the methods of analysis and the day-to-day problems faced by the working statistician.

REFERENCE

Hand, D. J., and Taylor, C. C. (1987). *Multivariate Analysis of Variance and Repeated Measures,* Chapman and Hall, London.

Assignments in Applied Statistics
Edited by S. Conrad
© 1989 John Wiley & Sons Ltd

Principal Components Analysis of Medical Records

Graham Dunn

Biometrics Unit, Institute of Psychiatry, University of London

OUTLINE

The aim of this assignment is to familiarize you with the use of principal components analysis (PCA) as an exploratory data reduction technique. The general aim is to produce a two- or three-dimensional display of a table of medical records. In this sense, the assignment will complement that on multidimensional scaling. The medical records are provided by 22 general practices, each practice contributing the number of records of each of several different types of psychiatric illness. PCA is an obvious technique to try to use for this type of data, but the method is not quite appropriate. You are encouraged to explore different approaches to the modification of the basic method to make it more suitable. In the supplementary questions you are encouraged to compare the use of PCA with that of factor analytic models.

KEYWORDS

principal components analysis, principal components, factor analysis, latent traits, cluster analysis, minimum spanning tree, distance measures, similarity coefficients, multidimensional scaling, correspondence analysis, multiple regression

Principal components analysis (PCA) is one of the most commonly used multivariate statistical techniques. Its major use is as an explanatory technique that can be used alone or in conjunction with other exploratory methods such as cluster analysis. PCA is essentially a method of data reduction which produces a few uncorrelated composite scores (principal components) which can then be used instead of the (possibly) large number of original measurements. Patterns of principal component scores can be explored using informal graphical techniques such as simple scatter diagrams, for example, or through the use of complex statistical procedures such as analysis of variance (ANOVA or MANOVA), discriminant analysis or multiple regression.

General descriptions of PCA can be found in textbooks of multivariate statistics such as Mardia, Kent and Bibby (1979, Ch. 8), and more practically orientated accounts can be found in Chatfield and Collins (1980, Ch. 4), Dunn and Everitt (1982, Ch. 4), Johnson and Wichern (1982, Ch. 8), Everitt and Dunn (1983, Ch. 4) and Dillon and Goldstein (1984, Ch. 2). Joliffe (1986) is a detailed monograph on the use of principal components analysis and similar techniques. The latter also includes an Appendix on the computation of principal components through the use of commercial computer packages such as BMDP, GENSTAT, MINITAB, SAS and SPSS[X]. Jeffers (1967) provides two case studies in the application of principal components analysis.

Typically, in the social and behavioural sciences, the first principal component score is used as a composite measurement of such concepts as educational attainment, cost of living or psychiatric distress, to give a few examples. When used in this way the results of PCA are superficially very similar to those obtained through the use of factor analysis models in the estimation of latent trait or factor scores. In fact, many non-statisticians refer to PCA as if it were a type of factor analysis. Similarly, many computer packages provide PCA as one of the many options in their factor analysis programs. The use of factor analysis models involves formal inferential techniques to fit hypothetical measurement models to sets of data. Like any other formal statistical method, factor analysis involves the reliance on several assumptions concerning the scales of measurement and distributional properties of the original variables. Principal components analysis, on the other hand, is simply a method of data transformation and does not rest on any detailed assumptions concerning the properties of the original measurements. You should be aware of the distinctions between the two methods both in their use of the techniques and in their interpretation of the work of others. For further details, see Everitt and Dunn (1983, Ch. 11) or Joliffe (1986, Ch. 7).

The present assignment is concerned with the use of PCA to examine patterns of medical records. There are several decisions to be made in carrying out a PCA of any data set. Should you analyse the covariance matrix for the original measurements or the correlation matrix? How many principal components (dimensions) are needed to describe the data? Are there other exploratory techniques which could be used in conjunction with PCA or even instead of PCA? Which method is more appropriate? Is there any evidence of clustering or of the presence of one or more outlying observations? Can the principal components or dimensions be interpreted? And so on.

THE DATA

The data to be analysed are provided in Table 1 (taken from Dunn, 1986). They are based on patterns of episodes of psychiatric disorders in patients registered with 22 general practices obtained from the longitudinal file of the Second National Morbidity Survey (Royal College of General Practitioners, 1980). This information was provided by the Medical Statistics Division of the Office of Population Censuses and Surveys. The file contains records of the number of episodes of psychiatric disorders (distinguished by a diagnostic code) for each of the six consecutive years of the survey (1970–6). In each episode there is at least one consultation with the general practitioner (GP) at which the relevant diagnosis has been made. For the purpose of the study described in Dunn (1986), the data were simplified in the following way. For each of the patients, for the whole of the six years of the study, it was asked whether they had experienced one or more episodes (equivalent to one or more consultations) of, say, depression. If the answer was 'yes' then the patient was coded as having a record of depression, but not otherwise. Similar codings were made for each of the other psychiatric disorders. To summarize, each patient contributes a single record of a particular disorder to Table 1 if, and only if, he or she has received a diagnosis of that disorder at least once during the period of the survey. Considering all psychiatric diagnoses, some patients will contribute no records at all, and others will contribute one, two or even more records. This latter aspect of the data might invalidate some forms of statistical analysis but, for the purpose of the present assignment, this complication will be ignored. It was also ignored in Dunn (1986).

Although the analysis of records of psychiatric illness might be regarded as a rather specialized aspect of applied statistics, the form of the data provided in Table 1 is typical of that frequently encountered in the biological, medical and social sciences. To verify that this is so, you need only refer to any book of tables of 'official statistics'.

Table 1

Distribution of records for each of the thirteen most common psychiatric disorders. (From Dunn, 1986, Table 1. *Reproduced by permission of Psychological Medicine*)

PRACTICE	PRACTICE SIZE[a]	TOTAL NO. PSYCHIATRIC PATIENTS[a]	RECORD TYPE[b]												
			130	134	150	135	146	147	126	136	148	132	131	125	139
A	2519	536	221	68	83	108	70	36	58	1	17	20	7	7	6
B	1504	444	281	184	11	3	69	17	0	2	9	1	1	7	0
C	2161	623	234	206	112	55	108	63	3	28	18	2	5	3	3
D	4187	1221	478	608	40	402	156	108	1	1	64	25	12	11	9
E	1480	493	76	251	173	159	23	37	4	30	14	1	9	6	3
F	2125	566	386	142	9	16	113	90	0	0	10	7	15	11	4
G	6514	2295	1122	742	197	435	81	172	330	193	66	33	22	16	21
H	1820	589	208	398	45	14	13	35	5	1	3	4	10	12	9
I	2671	1262	409	282	512	168	119	45	252	225	41	27	39	5	17
J	4220	903	314	429	105	53	65	37	20	1	29	28	23	14	11
K	2377	702	72	148	425	120	56	116	3	0	17	6	6	5	2
L	5009	1084	566	305	123	174	71	99	5	16	28	15	13	9	7
M	2037	430	241	207	4	4	37	29	0	1	11	0	3	0	0
N	1759	684	390	277	56	68	160	45	17	15	26	32	23	3	2
O	1767	573	248	178	61	89	122	70	13	11	16	19	5	6	3
P	3443	886	218	185	317	198	164	75	45	89	26	21	13	22	27
Q	2200	592	212	210	115	104	64	70	85	7	11	12	15	5	1
R	2639	756	280	155	286	180	53	74	36	4	27	9	16	3	12
S	1897	613	288	224	132	153	79	51	14	15	7	9	9	3	5
T	2278	846	331	251	170	439	76	84	9	24	20	8	5	5	7
U	2242	801	254	303	177	152	148	138	21	53	16	15	16	9	5
V	2497	852	330	290	186	215	119	121	21	7	16	26	13	13	9

[a] The number of patients registered with the practice throughout the whole of the six years of the Survey.

[b] Royal College Codes (in order of overall prevalence): 130, anxiety neurosis; 134, depressive neurosis; 150, unclassified symptoms; 135, physical disorders of presumably psychogenic origin; 146, insomnia; 147, tension headache; 126, affective psychosis; 136, neurasthenia; 148, enuresis; 132, phobic neurosis; 131, hysterical neurosis; 125, schizophrenia; 139, alcoholism and drug dependence.

Q1   Which data matrix should be analysed?

Start by considering a PCA of the record types (i.e. ignoring the first two columns of counts in Table 1). The results of a PCA will obviously be dependent on the choice of data matrix to be analysed. This should be demonstrated by separate analyses of (a) the covariance matrix for the record types and (b) the corresponding correlation matrix. In each case examine the pattern of eigenvalues obtained through the use of a scree plot (Everitt and Dunn, 1983, p. 46) to assess how well the first two principal components might represent the original data. Calculate principal component scores for each of the 22 practices and produce a two-dimensional scatter diagram using the first two principal components. How do these practices differ? Are there any unusual practices? Is there any evidence of clustering of practices? What do the first two principal components appear to be measuring?

You should consider carrying out a maximum likelihood factor analysis of the correlation matrix for the record types and compare the results with the corresponding PCA. Note that the terminology used in factor analysis is different to that used for PCA but, particularly if the PCA is carried out through an option in a factor analysis program, it is useful to be able to compare the results of a PCA and an equivalent factor analysis.

Q2   What is the influence of practice size?

Obviously the number of records of a particular type of psychiatric disorder is likely to be correlated with the size of the practice. Similarly, the number of records of a particular type might be correlated with overall levels of psychiatric morbidity (total number of psychiatric patients). Plot the principal component scores against either practice size or total number of psychiatric patients to look for any interesting associations.

A more challenging problem is to investigate the effect of practice size or levels of morbidity on the extraction of principal components. For each practice divide the number of records by either (a) practice size or (b) total number of psychiatric patients. Alternatively, each of the counts of the thirteen record types could be divided by their sum to provide a profile of relative frequencies. Investigate the effects of these modifications to the raw data matrix by repeating the analyses described in Question 1.

Q3   How can the PCA be supplemented by other methods of analysis?

Other forms of graphical representation of multivariate data are discussed in Joliffe (1986, Chs. 5 and 9) and Everitt and Dunn (1983,

Chs. 3–6). Consider how other methods might add to or be used to 'validate' the results of PCA and, in particular, investigate ways in which the results of cluster analyses (including minimum spanning trees—see Dunn, 1986) might be superimposed on scatter diagrams generated from principal components scores.

Q4   Are there 'better' methods of analysis?

Although it is stressed in statistics textbooks that PCA is simply a method of data transformation that is not dependent on detailed assumptions concerning the distributional properties of the observations, one still might suspect that data such as those given in Table 1 could be analysed in a more appropriate way than through an unmodified PCA. The aim of this question is to expose you to some of the alternative methods of analysis of tables of counts and proportions.

First, consider a table of 22 practice 'profiles'. The numbers in this table are the relative proportions within each practice of the thirteen diagnostic codes (see Table 3 of Dunn, 1986). Consider, for example, practices A and B with profiles

$$(P_{1A}, P_{2A}, \ldots, P_{iA}, \ldots, P_{13A})$$

and

$$(P_{1B}, P_{2B}, \ldots, P_{iB}, \ldots, P_{13B})$$

respectively. These profiles have already been subjected to a PCA in Question 2. Aitchison (1983) has suggested these profiles should be transformed prior to the PCA. He suggests the transformation (for practice A, for example)

$$x_{jA} = \log P_{jA} - \frac{1}{p} \sum_{i=1}^{p} \log P_{iA}$$

where p is the number of categories of record (13 in this case). Try this transformation and compare the results with those obtained for Question 2. Further details are provided in Joliffe (1986, pp. 209–12).

A second possibility is to define a suitable measure of similarity (distance) between profiles and then use some form of multi-dimensional scaling (see Dunn, 1986). One possibility for the similarity between profiles (say A and B) is given by

$$S_{AB} = \sum_{i=1}^{13} (P_{iA} P_{iB})^{1/2}$$

with corresponding distance given by

$$d_{AB} = (1 - S_{AB})^{1/2}$$

Calculate a distance matrix for the 22 practices in Table 1 and compare the results of multidimensional scaling of this distance matrix with the scatter diagrams produced for plotting principal components scores.

A third approach is through the use of correspondence analysis. This is a method of analysis which is particularly appropriate for tables of counts such as Table 1. The details of correspondence analysis are beyond the scope of this assignment but, basically, it is a form of multidimensional scaling of a distance matrix defined by

$$d_{AB}^* = \left[\sum_{i=1}^{13} \frac{1}{p_i} (P_{iA} - P_{iB})^2 \right]^{1/2}$$

where P_{iA} and P_{iB} are the profile proportions for code i of practices A and B respectively and p_i is the relative proportion of code i in the *whole* table (i.e. it is calculated using the column (code) totals of Table 1). A correspondence analysis of the counts in Table 1 will differ from most forms of multidimensional scaling in that (in the present context) each practice influences the final result through both its profiled shape *and* its overall 'mass' (i.e. the total number of records, which acts as a weighting factor for that practice in producing the solution). Large practices (those with large numbers of records) will influence the configuration of the solution more than small ones. Further details are provided in Greenacre (1984) and Lebart, Morineau and Warwick (1984). A GENSTAT program for correspondence analysis is listed in an Appendix in Greenacre (1984). Compare the results of a correspondence analysis of the data in Table 1 with those you have already obtained. For those readers without suitable software to carry out a correspondence analysis, a brief summary of the required results is provided in Dunn (1986).

REFERENCES

Aitchison, J. (1983). Principal component analysis of compositional data, *Biometrika*, **70**, 57–65.

Chatfield, C., and Collins, A. J. (1980). *Introduction to Multivariate Analysis*, Chapman and Hall, London.

Dillon, W. R., and Goldstein, M. (1984). *Multivariate Analysis*, Wiley, New York.

Dunn, G. (1986). Patterns of psychiatric diagnosis in general practice: the Second National Morbidity Survey, *Psychological Medicine*, 16, 573–81.

Dunn, G., and Everitt, B. S. (1982). *An Introduction to Mathematical Taxonomy*, Cambridge University Press, Cambridge.

Everitt, B. S., and Dunn, G. (1983). *Advanced Methods of Data Exploration and Modelling*, Gower Press, London.

Greenacre, M. J. (1984). *Theory and Applications of Correspondence Analysis*, Academic Press, London.

Jeffers, J. N. R. (1967). Two case studies in the application of principal component analysis, *Applied Statistics*, 16, 225–36.

Johnson, R. A., and Wichern, D. W. (1982). *Applied Multivariate Statistical Analysis*, Prentice-Hall, Englewood Cliffs, New Jersey.

Joliffe, I. T. (1986). *Principal Components Analysis*, Springer-Verlag, New York.

Lebart, L., Morineau, A., and Warwick, K. M. (1984). *Multivariate Descriptive Statistical Analysis*, Wiley, New York.

Mardia, K. V., Kent, J. T., and Bibby, J. M. (1979). *Multivariate Analysis*, Academic Press, London.

Royal College of General Practitioners (1980). Second National Morbidity Survey, *Journal of the Royal College of General Practitioners*, 30, 537–50.

SUPPLEMENTARY QUESTIONS

1. Suppose that a scientist has p instruments with which to measure a physical quantity X. He knows that each instrument measures a linear function of X and that all measurements are subject to independent random errors. Show that the problem of calibrating these instruments (i.e. investigating the relationships between the different instruments if the measurements were not subject to error) and of estimating their precisions is equivalent to fitting a one-factor model to a covariance matrix obtained from measuring a sample of objects with each of the p instruments. Demonstrate that this problem can only be solved if $p \geq 3$.

   Fit a one-factor model to the following covariance matrix for three measurements (X_1, X_2, X_3) on each of fifteen objects:

	X_1	X_2	X_3
X_1	1.98	2.27	2.58
X_2	2.27	2.48	2.84
X_3	2.58	2.84	3.02

2. The following (from Spearman, 1904) is a matrix of correlations between six examinations taken by 33 children:

	1	2	3	4	5	6
1 Classics	1.00	0.83	0.78	0.70	0.66	0.63
2 French	0.83	1.00	0.67	0.67	0.65	0.58
3 English	0.78	0.67	1.00	0.64	0.54	0.51
4 Mathematics	0.70	0.67	0.64	1.00	0.45	0.51
5 Discrimination	0.66	0.65	0.54	0.45	1.00	0.40
6 Music	0.63	0.58	0.51	0.51	0.40	1.00

Carry out a principal components analysis on these data and interpret the results. Compare this interpretation with that obtained from a maximum likelihood factor analysis.

3. Two standardized variables, X_1 and X_2, have a correlation matrix given by

$$R = \begin{bmatrix} 1 & r \\ r & 1 \end{bmatrix}$$

Show that the eigenvalues of R are $(1 + r)$ and $(1 - r)$, and find the principal components of R.

4. Explain the similarities and differences between factor analysis and principal components analysis.

5. Explain the differences between confirmatory factor analysis and exploratory factor analysis. Discuss the relative advantages and disadvantages of each.

6. Discuss the use and abuse of factor rotations. Illustrate your discussion with the results of both orthogonal and oblique rotations of the two-factor solutions obtained on answering Supplementary Question 2.

7. One of the major problems in the use of multiple regression in the social sciences is the potential multicollinearity of the independent variables. Discuss the use of principal components in the detection of collinearity and in the solution of this problem. (See Everitt and Dunn, 1983, Ch. 8; Joliffe, 1986, Ch. 8.)

FURTHER REFERENCE

Spearman, C. (1904). 'General intelligence', objectively determined and measured, *American Journal of Psychology*, **15**, 201–93.

Assignments in Applied Statistics
Edited by S. Conrad
© 1989 John Wiley & Sons Ltd

The Classification of Depression amongst Women

B. S. Everitt

Biometrics Unit, Institute of Psychiatry, University of London

OUTLINE

Cluster analysis techniques are exploratory tools which are used to search for groupings in initially unclassified data. They have been used in a variety of different areas, for example, market research, medicine and weather forecasting. This assignment involves their use in the development of diagnostic categories in psychiatric research. Such data pose particular problems, one of which is the mixture of variable types frequently used to describe each patient: the choice of similarity measure for such data is often critical. Each of these, and a number of more general problems, will need to be considered in this assignment.

KEYWORDS

cluster analysis, hierarchical clustering, similarity, standardization, missing values, classification of depression, assessing number of groups

Cluster analysis techniques which are used to generate classifications from initially unclassified material have found application in areas ranging from archaeology to market research; they have been used to produce classifications of amino acid sequences in proteins, survey respondents, consumer goods, geological data transmitted from earth satellites and even puberty rites of American Indian tribes. According to Wishart (1978) such classifications may help to

(a) formulate hypotheses concerning the origin of the population (e.g. in evolution studies);
(b) describe the sample in terms of a typology (e.g. for market analysis or administrative purposes);
(c) predict the future behaviour of population types (e.g. in modelling economic prospects for different industry sectors);
(d) optimize a functional process (e.g. search time in information retrieval);
(e) assist identification (e.g. in diagnosing diseases).

Useful general accounts of the methods are given by Hartigan (1975); Mardia, Kent and Bibby (1979, Ch. 13); Chatfield and Collins (1980, Ch. 11); Everitt (1980); Dillon and Goldstein (1984, Ch. 5); Johnson and Wichern (1982, Ch. 11); Hair, Anderson and Tatham (1987, Ch. 7); and in the excellent review paper by Cormack (1971).

One discipline which has been particularly attracted to the possibilities presented by numerical classification techniques is that of psychiatry, where many attempts have been made to use the methods to help refine or even redefine existing diagnostic categories and so produce a classification of the mentally ill which has more relevance for investigations of the causes and treatment of psychiatric disorders. Some examples of such studies are those described in Pilowsky, Levine and Boulton (1969), Everitt, Gourlay and Kendell (1971) and Paykel and Rassaby (1978).

The use of cluster analysis techniques in practice requires considerable care since there is a variety of problems that may require consideration. A number of these, choosing which method to use, deciding on the appropriate number of groups, etc., are relatively general (see Everitt, 1979); when using the methods on psychiatric data there are, however, often additional specific problems that have to be faced. Two of the most common are those of *missing data* and *mixed variable types*.

Missing data can arise for a number of reasons, failure to record a piece of information by the investigator, lack of recall on the part of the patient, etc.: the simplest way to deal with the problem is to drop those patients where it occurs. For some data sets, however, this might be very costly in loss of information and other approaches might be considered such as replacing the missing values for each variable by some appropriate 'typical' value calculated from the observations present for the variable.

When each patient is described by a number of variables which are *categorical*, others which are *ordinal* and yet others which purport to be *interval*, some thought needs to be given as to how to combine them all into a sensible measure of inter-individual *similarity* or *dissimilarity*, this being the basis of many methods of cluster analysis.

In this assignment both the general and the specific problems need to be considered.

THE DATA

The data for this assignment (see Table 1) consist of observations on eight variables for each of 118 female psychiatric patients. The variables recorded were as follows:

1. Age
2. IQ
3. Anxiety (measured on a four-point scale: none = 1, mild = 2, moderate = 3, severe = 4)
4. Depression (again measured on a four-point scale with the same categories as anxiety)
5. Can you sleep normally without tablets?
   (Yes = 1, No = 2)
6. Have you lost interest in sex?
   (No = 1, Yes = 2)
7. Have you thought recently about ending your life?
   (No = 1, Yes = 2)
8. Weight change over last six months (in pounds—negative values indicate a decrease)

The psychiatric investigator's interest lies in assessing whether there is any evidence that the data contain distinct subgroups of patients. Such subgroups would firstly provide a simple way to describe the patterns of similarities and differences among patients; this would be useful for communication. More excitingly, however, the groups might have implications for treatment and for research into the diagnosis of depressive disorders in women.

Q1   Which method of cluster analysis should be used?
     A number of studies have shown that it is unrealistic to expect a particular method of cluster analysis to be 'best' for all types of data in all situations. There are, however, results available from a number of empirical studies which may be of use in deciding between the

Table 1
Eight variable values for each of 118 female psychiatric patients

PATIENT	1. AGE	2. IQ	3. ANXIETY	4. DEPRESSION	5. SLEEP	6. SEX	7. LIFE	8. WEIGHT
1	39	94	2	2	2	2	2	4.9
2	41	89	2	2	2	2	2	2.2
3	42	83	3	3	3	2	2	4.0
4	30	99	2	2	2	2	2	−2.6
5	35	94	2	1	1	2	1	−0.3
6	44	90	M	1	2	1	1	0.9
7	31	94	2	2	M	2	2	−1.5
8	39	87	3	2	2	2	1	3.5
9	35	M	3	2	2	2	2	−1.2
10	33	92	2	2	2	2	2	0.8
11	38	92	2	1	1	1	1	−1.9
12	31	94	2	2	2	M	1	5.5
13	40	91	3	2	2	2	1	2.7
14	44	86	3	2	2	2	2	4.4
15	43	90	3	2	2	2	2	3.2
16	32	M	1	1	1	2	1	−1.5
17	32	91	1	2	2	M	2	−1.9
18	43	82	4	3	2	2	2	8.3
19	46	86	3	2	2	2	2	3.6
20	30	88	2	2	2	2	1	1.4
21	34	97	3	3	M	2	2	M
22	37	96	3	2	2	2	1	M
23	35	95	2	1	2	2	1	−1.0
24	45	87	2	2	2	2	2	6.5
25	35	103	2	2	2	2	1	−2.1
26	31	M	2	2	2	2	1	−0.4
27	32	91	2	2	2	2	1	−1.9

No.								
28	44	87	2	2	2	2	2	3.7
29	40	91	3	3	2	2	2	4.5
30	42	89	3	3	2	2	2	4.2
31	36	92	3	M	2	2	2	M
32	42	84	3	3	2	2	2	1.7
33	46	94	2	M	2	2	2	4.8
34	41	92	2	1	2	2	1	1.7
35	30	96	M	2	1	2	2	−3.0
36	39	96	2	2	2	1	1	0.8
37	40	86	2	3	2	2	2	1.5
38	42	92	3	2	2	2	1	1.3
39	35	102	2	2	2	2	2	3.0
40	31	82	2	2	2	2	1	1.0
41	33	92	3	3	2	2	2	1.5
42	43	90	M	M	1	2	2	3.4
43	37	92	2	1	2	1	1	M
44	32	88	4	2	2	2	M	M
45	34	98	2	2	2	2	2	0.6
46	34	93	3	2	2	2	1	0.6
47	42	90	2	1	1	1	1	3.3
48	41	91	3	1	1	2	1	4.8
49	31	M	3	2	2	2	1	−2.2
50	32	92	3	2	2	1	2	1.0
51	29	92	2	2	2	2	2	−1.2
52	41	91	2	2	2	2	2	4.0
53	39	91	2	2	2	1	2	5.9
54	41	86	2	1	1	2	1	0.2
55	34	95	2	1	1	1	1	3.5
56	39	91	1	1	2	1	1	2.9
57	35	96	3	2	2	1	1	−0.6
58	31	100	2	2	2	2	2	−0.6
59	32	99	4	3	2	2	2	−2.5

(continued overleaf)

Table 1 *Continued*

PATIENT	1. AGE	2. IQ	3. ANXIETY	4. DEPRESSION	5. SLEEP	6. SEX	7. LIFE	8. WEIGHT
60	41	89	2	1	2	1	1	.3.2
61	41	89	3	2	2	2	2	2.1
62	44	98	3	2	2	2	2	3.8
63	35	98	2	2	2	2	1	-2.4
64	41	103	2	2	2	2	2	-0.8
65	41	91	3	1	2	2	1	5.8
66	42	91	4	3	M	M	2	2.5
67	33	94	2	2	2	2	1	-1.8
68	41	91	2	1	2	2	1	4.3
69	43	85	2	2	2	1	1	M
70	37	92	1	1	2	2	1	1.0
71	36	96	3	3	2	2	2	3.5
72	44	90	2	M	2	2	2	3.3
73	42	87	2	2	2	1	2	-0.7
74	31	95	2	3	2	2	2	-1.6
75	29	95	3	3	2	2	2	-0.2
76	32	87	1	1	2	2	1	-3.7
77	35	95	2	2	2	2	2	3.8
78	42	88	1	1	1	2	1	-1.0
79	32	94	2	2	2	2	1	4.7
80	39	M	3	2	2	2	2	-4.9
81	34	M	3	M	2	2	1	M
82	34	87	3	3	2	2	1	2.2
83	42	92	1	1	2	1	1	5.0
84	43	86	M	3	2	2	2	0.4
85	31	93	M	2	2	2	2	-4.2
86	31	92	2	2	2	2	1	-1.1
87	36	106	2	2	2	1	2	-1.0
88	37	93	2	2	2	2	2	4.2

ID								
89	43	95	2	2	2	2	1	2.4
90	32	95	3	2	2	2	2	4.9
91	32	92	M	M	M	M	2	3.0
92	32	98	2	2	2	2	2	−0.3
93	43	92	2	2	2	2	1	1.2
94	41	88	2	2	2	2	1	2.6
95	43	85	1	1	1	1	1	1.9
96	39	92	2	2	2	2	2	3.5
97	41	84	2	2	2	2	1	−0.6
98	41	92	2	1	1	2	2	1.4
99	32	91	2	2	2	2	2	5.7
100	44	86	3	2	2	2	1	4.6
101	42	92	3	2	2	2	1	M
102	39	89	2	2	2	2	2	2.0
103	45	M	2	2	2	2	2	0.6
104	39	96	3	M	M	M	2	M
105	31	97	2	2	2	M	2	2.8
106	34	92	3	2	2	2	2	−2.1
107	41	92	3	2	2	2	2	−2.5
108	33	98	3	1	1	1	2	2.5
109	34	91	2	3	2	2	1	5.7
110	42	91	3	1	1	1	2	2.4
111	40	89	3	3	3	3	1	1.5
112	35	94	3	2	2	2	2	1.7
113	41	90	3	1	2	2	2	2.5
114	32	96	2	2	2	2	1	M
115	39	87	2	2	2	2	2	M
116	41	86	3	1	1	1	1	−1.0
117	33	89	1	1	1	1	1	6.5
118	42	M	3	2	2	2	2	4.9

M = missing value.

many competing methods. Examples of such studies are those described in Cunningham and Ogilvie (1972), Baker (1974), Milligan (1980) and Hands and Everitt (1987). Most of these investigations involve hierarchical clustering methods such as *single linkage*, *complete linkage*, etc., which are available in the CLUSTAN package (see Wishart, 1978). (For a general discussion of cluster analysis algorithms, see Späth, 1980.)

After studying the references given above select two or three hierarchical techniques with which to cluster the psychiatric data.

Q2 How should the missing data be dealt with?

Although the psychiatric data in Table 1 does contain a number of missing values, the problem is not as serious as for many data sets since the majority of observations are complete. Nevertheless, it has to be decided whether to simply drop observations containing one or more missing values or replace missing values with some appropriate 'typical' value. In the latter case some thought needs to be given to what would be a sensible replacement value to use for the different types of variable.

After deciding on the appropriate 'typical' values for the different variable types, investigate whether the two proposed strategies for dealing with the missing values in the psychiatric data affect the clustering solutions.

(A detailed account of methods for dealing with missing data is given in Little and Rubin, 1987.)

Q3 How should inter-individual similarity or dissimilarity be measured?

The eight variables describing each psychiatric patient consist of a number of categorical variables (numbers 5, 6 and 7), some ordinal variables (numbers 3 and 4) and finally three interval level variables (numbers 1, 2 and 8, although the measurement status of IQ is, perhaps, ambivalent). Before using most methods of clustering (and certainly before using any of the hierarchical techniques), a measure of inter-individual similarity or dissimilarity must be constructed from the variable values of each pair of individuals. One of the most common such measures is Euclidean distance, defined as

$$d_{ij} = \left[\sum_{k=1}^{p} (x_{ik} - x_{jk})^2 \right]^{1/2} \qquad (1)$$

where $x_{ik}, x_{jk}, k = 1, \ldots, p$, represent the values taken by individuals i and j on the p variables observed.

It is clear that using this measure on the raw psychiatric data would not be particularly sensible (why?) and alternative measures need to be considered. One possibility would be to standardize the observations in some way before calculating the Euclidean distance. Standardization via the standard deviations calculated from the whole data set is often used, with Euclidean distances then being calculated not from the raw data values, x_{ij}, but from transformed values given by

$$z_{ij} = \frac{x_{ij}}{s_j} \tag{2}$$

where s_j is the standard deviation of variable j. Such a procedure is often used, although it is not completely satisfactory (see Fleiss and Zubin, 1969); certainly it seems of dubious validity for the psychiatric data with its mixture of variable types.

An alternative to the Euclidean distance and one which should be more suited to data containing variables of different types is the similarity measure suggested by Gower (1971). This coefficient is defined as follows:

$$s_{ij} = \frac{\sum\limits_{k=1}^{p} s_{ijk}}{\sum\limits_{k=1}^{p} w_{ijk}} \tag{3}$$

The weights, w_{ijk}, are set equal to unity or zero depending on whether the comparison of individuals i and j on variable k is considered valid or not. If the value of variable k is missing for one or both the individuals then w_{ijk} is set equal to zero. The terms in the numerator of (3) are defined differently for categorical and continuous variables:

(a) *Categorical.* In this case $s_{ijk} = 1$ if the two individuals i and j take the same value on variable k and $s_{ijk} = 0$ otherwise.
(b) *Continuous.* Here s_{ijk} is given by

$$s_{ijk} = 1 - \frac{|x_{ik} - x_{jk}|}{R_k} \tag{4}$$

where x_{ik} and x_{jk} are the values taken by individuals i and j on variable k and R_k is the range of variable k in the sample.

(In both cases (a) and (b), s_{ijk} takes the value zero when w_{ijk} equals zero.)

Compare the results obtained from your chosen clustering technique when applied to:

(a) Euclidean distance calculated from the raw data,
(b) Euclidean distance calculated from the standardized data,
(c) Gower's similarity coefficient.

Q4 How might the number of groups in the data be determined?

One of the most difficult problems encountered when using clustering techniques in practice is determining the number of groups or clusters which 'best' fit the data in some sense. The problem has been discussed in detail by a number of authors including Friedman and Rubin (1967), Beale (1969) and Everitt (1979), and it is clear that no completely satisfactory solution is available.

When using one or other of the hierarchical clustering techniques a suggestion often made is to plot the number of groups against linkage level and to examine the resulting plot for an 'elbow' or sharp change in going from $i-1$ to i groups; such a change is generally taken as indicating the presence of i groups in the data. (Figure 1 shows what such a plot might look like.)

One of the obvious weaknesses with the above procedure is the degree of subjectivity likely to be involved. Mojena (1977) has attempted to overcome this problem by introducing a rule based on the values of the $n-1$ fusion levels in the hierarchy (where n is the number of individuals to be classified), and selects the partition corresponding to the first stage, j, in the cluster sequence $j = 2, \ldots, n-2$, satisfying

$$Z_{j+1} > \bar{Z} + k s_z \qquad (5)$$

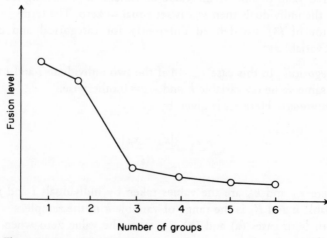

Figure 1
Hypothetical example of plot of fusion level against number of groups

where $\bar{Z}$ and s_z are the mean and unbiased standard deviation of the distribution of the $n-1$ values and k is an appropriate Standard Normal deviate (for example, 1.96). This procedure is implemented in the CLUSTAN package.

Investigate the use of the above procedures for determining the number of groups on each of the clustering solutions produced by using different methods of cluster analysis and different measures of similarity or dissimilarity.

OTHER POSSIBLE ANALYSES

One of the most interesting methods of cluster analysis, at least from a statistical point of view, is that based upon fitting *mixtures of multivariate Normal distributions*: the procedure is described in detail in Wolfe (1970) and Everitt and Hand (1981). The technique is clearly not suitable for the raw psychiatric data with its mixture of variable types, but it might be suitable for the data after transforming to principal component scores.

Questions that need consideration are:

(a) Should the principal components analysis be performed on the covariance or correlation matrix?
(b) How many principal component scores should be used in the mixture analysis?
(c) Should the anxiety and depression scores be used directly or replaced by a series of dummy variables?

The principal components analysis may also be useful for giving a graphical representation of the data, of use in interpreting the clustering results.

Q5 Plot the first few principal component scores against each other to produce a series of two-dimensional representations of the psychiatric data.

REFERENCES

Baker, F. B. (1974). Stability of two hierarchical grouping techniques. Case 1: sensitivity to data errors, *Journal of the American Statistical Association*, **69**, 440–5.

Beale, E. M. L. (1969). Euclidean cluster analysis, *Bulletin of the International Statistical Institute*, **43**, 92–4.

Chatfield, C., and Collins, A. J. (1980). *Introduction to Multivariate Analysis*, Chapman and Hall, London.

Cormack, R. M. (1971). A review of classification (with discussion), *Journal of the Royal Statistical Society (Series A)*, **134**, 321–67.

Cunningham, K. M., and Ogilvie, J. C. (1972). Evaluation of hierarchical grouping techniques: a preliminary study, *The Computer Journal*, **15**, 209–13.

Dillon, W. R., and Goldstein, M. (1984). *Multivariate Analysis*, Wiley, New York.

Everitt, B. S. (1979). Unresolved problems in cluster analysis, *Biometrics*, **35**, 169–81.

Everitt, B. S. (1980). *Cluster Analysis*, Heinemann, London.

Everitt, B. S., Gourlay, A. J., and Kendell, R. E. (1971). An attempt at validation of traditional psychiatric syndromes by cluster analysis, *British Journal of Psychiatry*, **119**, 399–412.

Everitt, B. S., and Hand, D. J. (1981). *Finite Mixture Distributions*, Chapman and Hall, London.

Fleiss, J. L., and Zubin, J. (1969). On the methods and theory of clustering, *Multivariate Behavioral Research*, **4**, 235–50.

Friedman, H. P., and Rubin, J. (1967). On some invariant criteria for grouping data, *Journal of the American Statistical Association*, **62**, 1159–78.

Gower, J. C. (1971). A general coefficient of similarity and some of its properties, *Biometrics*, **27**, 857–72.

Hair, J. F., Anderson, R. W., and Tatham, R. L. (1987). *Multivariate Data Analysis*, Macmillan, New York.

Hands, S., and Everitt, B. (1987). A Monte Carlo study of the recovery of cluster structure in binary data by hierarchical clustering techniques, *Multivariate Behavioral Research*, **22**, 235–43.

Hartigan, J. A. (1975). *Clustering Algorithms*, Wiley, New York.

Johnson, R. A., and Wichern, D. W. (1982). *Applied Multivariate Statistical Analysis*, Prentice-Hall, Englewood Cliffs, New Jersey.

Little, R. J. A., and Rubin, D. B. (1987). *Statistical Analysis with Missing Data*, Wiley, New York.

Mardia, K. V., Kent, J. T., and Bibby, J. M. (1979). *Multivariate Analysis*, Academic Press, London.

Milligan, G. W. (1980). An examination of the effect of six types of error perturbation of fifteen clustering algorithms, *Psychometrika*, **45**, 325–42.

Mojena, R. (1977). Hierarchical grouping methods and stopping rules: an evaluation, *The Computer Journal*, **20**, 359–63.

Paykel, E. S., and Rassaby, E. (1978). Classification of suicide attempters by cluster analysis, *British Journal of Psychiatry*, **133**, 45–52.

Pilowsky, I., Levine, S., and Boulton, D. M. (1969). The classification of depression by numerical taxonomy, *British Journal of Psychiatry*, **115**, 137–45.

Späth, H. (1980). *Cluster Analysis Algorithms*, Ellis Horwood, Chichester.

Wishart, D. (1978). *CLUSTAN User Manual*, University of Edinburgh.

Wolfe, J. H. (1970). Pattern clustering by multivariate mixture analysis, *Multivariate Behavioral Research*, **5**, 329–50.

SUPPLEMENTARY QUESTIONS

1. There are six well-known hierarchical clustering techniques:
   (a) single linkage,
   (b) complete linkage,
   (c) centroid,
   (d) median,
   (e) group average,
   (f) Ward's method.

   Which of these are invariant to monotonic transformations of the inter-individual similarity or dissimilarity matrix? Would you consider this invariance property to be important and why?

2. The multivariate mixture approach to clustering is clearly not suitable for data consisting of categorical, ordinal and interval scaled variables. Suggest how the model might be used for such data by assuming that the values of the categorical and ordinal variables are generated from an underlying continuous variable by applying a threshold. (See Everitt, 1988.)

3. Paykel and Rassaby (1978), in an investigation of suicide attempters, produced a three-group classification using Ward's method of hierarchical clustering. They then carried out a one-way analysis of variance to assess which of the variables used to produce the classification differed significantly between groups. Why are the usual F-tests invalid in these circumstances?

4. Many of the problems of cluster analysis arise because of the difficulty of defining the term 'cluster' in any very generally acceptable manner. Make some intuitively sensible suggestions and comment on why they may not be appropriate in all situations. (See Everitt, 1980, and Everitt and Dunn, 1983, for a detailed account of the problem of definition.)

5. Many psychiatrists have expressed reservations about the worth of *any* classification of the mentally ill, arguing that labelling patients in this way presents too simplistic a view of psychiatric disorders. What would you see as the main reasons for seeking such a classification and how would you judge its usefulness and validity?

6. What are the advantages and disadvantages of methods of cluster analysis which produce hierarchical classifications compared with those that produce non-hierarchical solutions? Which do you think would be more suitable for psychiatric data and why? In what areas would hierarchical classifications clearly be the most appropriate?

7. Apart from principal components analysis, what other methods might be used to produce a display that might be useful in the interpretation of the results of a cluster analysis?

8. Cluster analysis and factor analysis are often confused by research workers, particularly in psychology. Compare and contrast the two types of analysis, giving examples of situations in which one is more suitable than the other.

FURTHER REFERENCES

Everitt, B. S., and Dunn, G. (1983). *Advanced Methods of Data Exploration and Modelling*, Gower Press, London.

Everitt, B. S. (1988). A finite mixture model for the clustering of mixed-mode data, *Statistics and Probability Letters*, **6**, 305–9.

Assignments in Applied Statistics
Edited by S. Conrad
© 1989 John Wiley & Sons Ltd

Discriminant Analysis for Psychiatric Screening

D. J. Hand

Faculty of Mathematics, The Open University

OUTLINE

Discriminant analysis techniques are statistical methods for characterizing the differences between groups and for defining classification rules for assigning objects to classes based on measurements taken on those objects. In this assignment the data are responses to 25 questions taken on each of 60 subjects, who each fall into one of three groups according to their degree of psychiatric illness. The basic objective is to formulate a classification rule for diagnosing the severity of illness of future patients, again based on their responses to the 25 questions.

You are expected to deal with missing data, overfitting and variable selection, to obtain a reliable estimate of future performance of the derived classification rule and to see how the ordinal nature of the responses to the questions affects the results.

KEYWORDS

discriminant analysis, allocation, classification, variable selection, error rate, stepwise methods, leaving-one-out method, jackknife, bootstrap, missing values, ordinal data

Discriminant analysis techniques are statistical methods used for characterizing the differences between groups of objects. They are used in two main ways. The first way is to present a description of the aspects of the objects which are important in distinguishing between the groups, perhaps combining these aspects into some simple mathematical function on which the groups are maximally different. The second way has as its aim the production of a rule which can be used to classify future objects, for which the true classification may not be known but for which the aspects and characteristics can be measured.

There are many different approaches to discriminant analysis. Overviews are presented by Mardia, Kent and Bibby (1979, Ch. 11); Hand (1981); Devijver and Kittler (1982); Johnson and Wichern (1982, Ch. 10); Dillon and Goldstein (1984, Chs. 10 and 11), and Hair, Anderson and Tatham (1987, Ch. 3). The oldest and most widely implemented method in statistical packages is that originated by Fisher (1936). This is based on descriptions of the groups in terms of first- and second-order moments of the aspects and variables describing the objects. This classical approach is described in depth in Lachenbruch (1975) and Klecka (1980). More recent developments include the kernel method (Hand, 1982) and nearest neighbour methods (Devijver and Kittler, 1982).

Because of the widespread availability of the classical method, this assignment will be based on the assumption that the reader will use that method. References to possible explorations using other methods will be restricted to the supplementary questions.

One of the first issues one must address in a typical discriminant analysis is that of selecting the variables to be used to describe the differences between groups. If the aim is simply to characterize the between-groups differences then selecting the variables is the fundamental task. If, however, the aim is to produce a classification rule, why might one want to select a subset of variables? Surely extra variables can only *add* information about the differences between the groups, and never subtract information, so that using more variables must inevitably lead to improved classification of future objects.

In fact this is not quite true. As more variables are used and a more complicated classification function based on them is produced, so there is an increasing danger of overfitting. Thus one gets very good classification results on the data being used to estimate the classification rule (the 'design set') but poor generalization to new objects.

There are a number of different methods available for selecting variables. The overall principle is to choose a measure of separability between the groups and then either sequentially accumulate the variables which maximize this measure of separation or, beginning with all available variables, sequentially eliminate those variables whose removal leads to least reduction in separability. Such *stepwise* variable selection methods are widely used in some application areas of statistics.

Blind application of such methods is to be discouraged, especially when the objective is the description of the ways in which the groups differ. Such description is intended to serve as an aid to understanding—to which automatic data analysis is the antithesis.

A further cautionary note about variable selection applies to the F-values, reported by many packages, indicating the change in separation between groups when each variable is added or deleted. Such F-values can be interpreted as measures of separation, but they cannot be used as the basis of significance tests by referring them to standard F-tables. Because the variables are selected by *maximizing* or *minimizing* the change in separation resulting from adding or deleting them, the reported F-statistics do not follow standard F-distributions.

Excellent reviews of variable selection methods are given in McKay and Campbell (1982a, 1982b).

If classification is the ultimate aim, one will be interested in an estimate of future classification performance; that is one will want to know the likely proportion of future cases which will be correctly or incorrectly classified (this latter is the *misclassification* or *error rate*).

There are many different ways in which error rate can be estimated. A review is given in Hand (1986). An early method is the *resubstitution* method in which the objects used to design the classification rule are reclassified by the rule to determine the proportion misclassified. For obvious reasons this leads to optimistically biased results (see, for example, Hand, 1983). Several discriminant analysis packages, however, still report it.

A more recent method is the *leaving-one-out* (sometimes called cross-validation) method which involves recalculating the classification rule as many times as there are data points. Details are given in Hand (1986).

A further method which is superficially similar is the *jackknife* method. In fact, this has a completely different theoretical basis. The output of some packages, however, has annotations which confuse the two methods.

Efron (1982) describes these methods, and also describes the *bootstrap* method (see also Efron, 1983) and extensions of it. This latter is a very exciting development which holds great promise for the future.

The problem of overfitting arising from too many variables for the available number of data points and the problem of getting an accurate estimate of future performance from only a limited amount of data are problems which have received great attention in the discriminant analysis literature and which have parallels in other areas of statistics. Other problems which can cause difficulty are the treatment of ordinal, categorical and mixed data (see, for example, Goldstein and Dillon, 1978;

Krzanowski, 1975) and how to handle missing values (see Dempster, Laird and Rubin, 1977).

THE DATA

The data consist of 25 subjective scores of feelings of inadequacy, tension, etc., extracted from the 140 items described in Goldberg (1972). Twenty subjects from each of three groups scored each item. The groups were normal people, people diagnosed by psychiatrists as mildly ill and people diagnosed by psychiatrists as severely ill. These are coded 1, 2 and 3

Table 1
The 25 items

1. Have you recently felt that you are playing a useful part in things?
2. Have you recently felt contented with your lot?
3. Have you recently felt capable of making decisions about things?
4. Have you recently felt that you are just not able to make a start on anything?
5. Have you recently felt yourself dreading everything that you have to do?
6. Have you recently felt constantly under strain?
7. Have you recently felt you could not overcome your difficulties?
8. Have you recently felt afraid of expressing yourself in case you made a foolish mistake?
9. Have you recently felt frightened to be on your own?
10. Have you recently felt confident about going into public places?
11. Have you recently felt afraid to read the papers or watch TV because of what you might see?
12. Have you recently felt as though you were not really there or as though things around you were not real?
13. Have you recently been finding life a struggle all the time?
14. Have you recently been blaming yourself for things that have gone wrong?
15. Have you recently been able to enjoy your normal day-to-day activities?
16. Have you recently been taking things hard?
17. Have you recently been getting edgy and bad tempered?
18. Have you recently been getting scared or panicky for no good reason?
19. Have you recently been able to face up to your problems?
20. Have you recently been having a lot of worry about money?
21. Have you recently found everything getting on top of you?
22. Have you recently found yourself getting easily upset about things?
23. Have you recently found little annoyances making you upset or angry?
24. Have you recently had the feeling that people were looking at you?
25. Have you recently noticed that your feelings are easily hurt?

respectively. Each score is coded, in order of increasing severity, 1, 2, 3 or 4, with 0 signifying a missing value. Details of the items are given in Table 1. The data appear in Table 2.

The objective is to derive a classification rule which can be used to classify future subjects into one of the three classes on the basis of their scores on the 25 items.

Q1 How should the missing data be dealt with?

Two common approaches to missing data are to drop cases which have missing values from the analysis altogether and to impute substitute values for missing ones. The first approach is feasible provided it does not lead to too drastic a loss of data and provided that dropping incomplete cases does not distort the samples in some way. Thus before adopting this approach one should examine the pattern of missing data to see if the samples are likely to be distorted. (Will, for example, all the less severely ill subjects in group 3 be dropped using this approach?)

Imputation can be very simple (e.g. use the mean for that group on the relevant variable) or very complicated (e.g. iterative regression, updating the values in each variable one variable at a time).

In statistical techniques based on first- and second-order statistics one can calculate means and covariance matrices using all the relevant data values which are present. Some packages have this facility. There is, however, a danger that the resulting matrices will not be positive definite.

Little and Rubin (1987) review methods for handling missing data.

What effects do these various methods have on the results of the analyses below?

Q2 Is the dimensionality to sample size ratio a problem?

In problems with many variables and relatively few sample points there is a danger of overfitting—of producing a result that fits the given data very well but which generalizes to future data very poorly. *Ad hoc* rules of thumb suggest that one should have at least five or ten times as many data points as cases to avoid this—though the ratio also depends on the distributions and the nature of the data involved.

Since here we have 25 variables and only 60 subjects it is natural that this question should be of concern.

To see if it might be important calculate the simple resubstitution error rate for the 60 cases on all 25 variables and compare it with the estimate of error rate obtained by some other method (such as the leaving-one-out method). A wildly optimistic resubstitution estimate suggests that overfitting is a problem.

Table 2
Discriminant analysis data

CLASS (1 = HEALTHY, (2 = MILD ILLNESS, 3 = SEVERE ILLNESS)	CASE NUMBER WITHIN CLASS	ITEMS NUMBER 1 TO 25 IN ORDER
1	1	2222232122211212222222222
1	2	2221222212111211222221222
1	3	1121233322211211333331233
1	4	2221222111111211122121221
1	5	1121222111111211112120000
1	6	1121122112111211111221222
1	7	2222222222221212232322232
1	8	1121222111111211232211232
1	9	1121222111111211112111221
1	10	2122222111111211222121232
1	11	2221232212121211222232222
1	12	2121222112111211222121221
1	13	1122232113211210112121232
1	14	1121222111111211112121221
1	15	3323341222231212232232223
1	16	2222242211121211112211222
1	17	1111111111111111111111121
1	18	3121232123221211233222233
1	19	1221222121221211112122222
1	20	1121232113211211331222112
2	1	2112221122211211222222222
2	2	3121212212121211212121241
2	3	3223222222223211332333300
2	4	2122222123121213222211221
2	5	2321232111211333232232333
2	6	1121222312311211223331212
2	7	2121123112222221222212212
2	8	1121322121111211121112211
2	9	2222222211111221122111211
2	10	3134444444441400000000000
2	11	2221212112121121111121121
2	12	3222232232213343322333223
2	13	3223322212221211122232213
2	14	1222111213221212211111131
2	15	2243342223421212323343343
2	16	3222330233331213033333323
2	17	4433343334334413333344444
2	18	3332222123432212323233322
2	19	4244443344423143434434444
2	20	2122222222211211332322232

Table 2   (*Continued*)

CLASS (1 = HEALTHY, 2 = MILD ILLNESS, 3 = SEVERE ILLNESS)	CASE NUMBER WITHIN CLASS	ITEMS NUMBER 1 TO 25 IN ORDER
3	1	3131332113221131123331321
3	2	1121221211111211112111111
3	3	3324333334441213444424424
3	4	2323433333332223333333333
3	5	3334343433433232344433323
3	6	2121342314231212434422343
3	7	4444444444443211440133444
3	8	2123343222322211332232322
3	9	4414444444444434444424414
3	10	4123444244441212324044444
3	11	3322330213333231334333333
3	12	1121222211111211112111211
3	13	4121231133112101323312310
3	14	3323332222233222333332333
3	15	1122222221122211122222211
3	16	4323243224332341244444314
3	17	3121222313331311334433413
3	18	3244443434432211444414444
3	19	1200244214434444424234400
3	20	4444444444441233344444443

Q3 How shall we select a good subset of variables?

First we must choose a separability criterion so that we can identify whether one set of variables is better than another or not. If our ultimate aim is to produce a small error rate then ideally we would use error rate as the separability criterion. This, however, is computationally intensive and thus other separability measures are usually used. Common ones include Wilks's lambda, Mahalanobis's distance and the smallest F-ratio between pairs of groups.

Having chosen a criterion we must also decide whether to use forward or backward stepwise methods of variable selection or some more complicated method.

Compare the effects of different separability criteria and different selection methods in terms of the variable sets produced. Plot the relationship of estimated error rate against dimensionality, using both resubstitution and some other error rate estimation method to see if classification performance does deteriorate beyond a certain number of variables.

Q4   Does the ordinal nature of the data affect things?
Classical discriminant analysis, based on first- and second-order moments, assumes interval scale data. The items here, however, are only scored on an ordinal scale. It is possible that some arbitrary monotonic transformation, perfectly legitimate in terms of the items themselves, might lead to substantially different discriminant analysis results.
Try various monotonic transformations of the items to see how the discriminant analysis results are affected.

REFERENCES

Dempster, A. P., Laird, N. M., and Rubin, D. B. (1977). Maximum likelihood from incomplete data via the EM algorithm (with discussion), *Journal of the Royal Statistical Society (Series B)*, 39, 1–38.

Devijver, P. A., and Kittler, J. (1982). *Pattern Recognition: A Statistical Approach*, Prentice-Hall, Englewood Cliffs, New Jersey.

Dillon, W. R., and Goldstein, M. (1984). *Multivariate Analysis*, Wiley, New York.

Efron, B. (1982). *The Jackknife, the Bootstrap and Other Resampling Plans*, Society for Industrial and Applied Mathematics, Philadelphia, Pennsylvania.

Efron, B. (1983). Estimating the error rate of a prediction rule: improvement on cross-validation, *Journal of the American Statistical Association*, 78, 316–31.

Fisher, R. A. (1936). The use of multiple measurements in taxonomic problems, *Annals of Eugenics*, 7, 179–88.

Goldberg, D. P. (1972). *The Detection of Psychiatric Illness by Questionnaire*, Oxford University Press, Oxford.

Goldstein, M., and Dillon, W. R. (1978). *Discrete Discriminant Analysis*, Wiley, New York.

Hair, J. F., Anderson, R. W., and Tatham, R. L. (1987). *Multivariate Data Analysis*, Macmillan, New York.

Hand, D. J. (1981). *Discrimination and Classification*, Wiley, Chichester.

Hand, D. J. (1982). *Kernel Discriminant Analysis*, Research Studies Press, Letchworth.

Hand, D. J. (1983). A comparison of two methods of discriminant analysis applied to binary data, *Biometrics*, 39, 683–94.

Hand, D. J. (1986). Recent advances in error rate estimation, *Pattern Recognition Letters*, 4, 335–46.

Johnson, R. A., and Wichern, D. W. (1982). *Applied Multivariate Statistical Analysis*, Prentice-Hall, Englewood Cliffs, New Jersey.

Klecka, W. R. (1980). *Discriminant Analysis*, Sage Publications, Beverly Hills, California.

Krzanowski, W. J. (1975). Discrimination and classification using both binary and continuous variables, *Journal of the American Statistical Association*, 70, 782–90.

Lachenbruch, P. A. (1975). *Discriminant Analysis*, Hafner Press, New York.

Little, R. J. A., and Rubin, D. B. (1987). *Statistical Analysis with Missing Data*, Wiley, New York.

Mardia, K. V., Kent, J. T., and Bibby, J. M. (1979). *Multivariate Analysis*, Academic Press, London.

McKay, R. J., and Campbell, N. A. (1982a). Variable selection techniques in discriminant analysis. I: Description, *British Journal of Mathematical and Statistical Psychology*, **35**, 1–29.

McKay, R. J., and Campbell, N. A. (1982b). Variable selection techniques in discriminant analysis. II: Allocation, *British Journal of Mathematical and Statistical Psychology*, **35**, 30–41.

SUPPLEMENTARY QUESTIONS

1. The classical method of discriminant analysis assumes that the classes have identical covariance matrices. Given the categorical nature of the data, taking only four possible values, this may not be a realistic assumption. Some packages permit classification results to be produced, relaxing this assumption. Sometimes, however, the potential benefits are counterbalanced by the increase in the number of parameters that need to be estimated. Explore this, expressing the final results in terms of error rate estimated both by resubstitution and by some other method such as leaving-one-out.

2. Classical methods, based on first- and second-order moments, cannot take account of small but important fluctuations in the probability distributions concerned. Non-parametric methods can do this, but there is a risk of overfitting. Compare the use of such a non-parametric method (such as the nearest-neighbour routine in SAS) with the classical method. How is the comparison influenced by the number of variables involved?

   Nearest-neighbour methods, like the classical method, in fact assume more than ordinal levels of measurement for the items. Try recoding the items to binary (1 and 2 going to 0; 3 and 4 going to 1) and seeing how this affects the results.

3. Taking a pair of groups, compare the performance of logistic regression with the classical approach.

4. Another approach to variable selection takes advantage of the researcher's knowledge of what the items mean by requiring the researcher to group the items into subscales. The items in each subscale are then combined to produce a single score for the domain spanned by the constituent items. A very simple way to produce such a score would be to add the subscale items. One possible advantage of this kind of procedure is that the resulting subscale scores might have statistical properties better suited to the application of classical discriminant analysis. Explore this possibility, comparing the results with the direct applications of discriminant analysis above.

5. Classification performance of a rule can sometimes be improved by using 'partial discriminant analysis' or 'the reject option'. These are names for the practice of not classifying all new cases into the defined groups but of permitting cases about which one is uncertain to be rejected from the classification procedure altogether (perhaps so that further information can be collected on them). Explore this strategy for the data in this assignment using, as the basis for the rejection criterion the probabilities of belonging to each class, output from a standard discriminant analysis program.

6. Explore the possibility of developing two different discriminant functions, one for distinguishing between normals and those who are ill (whether mildly or severely) and the second for distinguishing the mildly from the severely ill.

Assignments in Applied Statistics
Edited by S. Conrad
© 1989 John Wiley & Sons Ltd

Understanding Pain through Multidimensional Scaling

B. S. Everitt

Biometrics Unit, Institute of Psychiatry, University of London

OUTLINE

Proximity matrices arising from experiments in which subjects are asked to assess the similarity of pairs of stimuli are frequently analysed by representing them in the form of a simple geometrical model or picture. The class of techniques involved is known as multidimensional scaling, and particular members of the class have been used to analyse the perceived judgements of similarity of spectral colours, countries, etc.

In applying the methods in practice, choices have to be made between the different techniques and between solutions of differing dimensionality. Consideration also needs to be given to the substantive interpretation of the derived geometrical configuration. In this assignment, concerned with similarity of adjectives describing pain, each of these problems is of importance.

KEYWORDS

multidimensional scaling, metric and non-metric scaling, assessment of pain, MULTISCAL, local optimum, similarity

A frequently encountered type of data particularly in the behavioural sciences is the *proximity matrix*, arising either directly from experiments in which subjects are asked to assess the similarity (or in some cases the dissimilarity) of two stimuli or indirectly as a measure of the correlation or covariance of the two stimuli deriving from a set of measurements describing each. The most common class of techniques used to analyse such data is that of *multidimensional scaling*. The underlying purpose that members of this class share, despite their apparent diversity, is to represent the structure or pattern in the proximity matrix by a simple geometrical model or picture. Useful general accounts of multidimensional scaling techniques are given in Shepard (1974); Kruskal and Wish (1978); Mardia, Kent and Bibby (1979, Ch. 14); Chatfield and Collins (1980, Ch. 10); Dillon and Goldstein (1984, Ch. 4); Hair, Anderson and Tatham (1987, Ch. 8).

Multidimensional scaling has proved particularly attractive to psychologists attempting to extract the underlying dimensions of similarity judgements made on a wide variety of types of stimuli. Examples include spectral colours (Ekman, 1954; Shepard, 1962), musical intervals (Levelt, Van de Geer and Plomp, 1966), countries (Wish, 1971; Wish, Deutsch and Biener, 1972), and societal problems (Carroll and Wish, 1974). Detailed descriptions of a number of other applications of multidimensional scaling are given in Romney, Shepard and Nerlove (1972).

As with most complex methods of analysis the user of multidimensional scaling techniques is faced with a variety of problems. These include the choice of an appropriate method, particularly whether it should be *metric* or *non-metric*, how to choose the proper number of dimensions in which to represent the observed similarities and the problem of achieving a sensible substantive interpretation of the derived geometrical configuration.

THE PROBLEM—DESCRIBING PAIN

In medical research and practice, assessing and understanding the nature of a patient's pain complaint is an important and difficult clinical task. Reaching the correct diagnosis and thereby selecting the appropriate treatment may depend on an accurate assessment of the precise characteristics of the pain.

As part of a study to construct a measuring instrument to assess pain, a psychologist was interested in people's judgement of the similarity between a number of adjectives that might be used to describe a particular pain quality. An experiment was performed in which each of the adjectives was transcribed on to a separate card, and subjects were then asked to sort the cards into groups describing a similar pain quality. (The example, cool, cold, freezing, was given as indicating differing intensities of a pain sensation or quality.)

The measure of inter-adjective similarity actually used in the original study was one derived from principles of information theory. It is described in detail in Burton (1972), but basically involves the probabilities that a subject i will place two words in the same group or in different groups, given that we have no information about which words have been placed in the group. These probabilities are given by

$$P_{ij} = \frac{N_{ij}(N_{ij}-1)}{N_w(N_w-1)}, \qquad Q_i = 1.0 - \sum_{j=1}^{N_i} P_{ij} \qquad (1)$$

where $P_{ij} = p$ (subject i places two words in the same group, j)

$Q_i = p$ (subject i places the words in different groups)

N_{ij} = number of words placed in group j by subject i

N_i = number of groups given by subject i

N_w = total number of words

The 'information content' of the events 'two words are in the same group' and 'two words are in different groups' for this particular subject is given by the negative of the logarithm to base two of P_{ij} and Q_i respectively (see, for example, Siegel, 1956).

The similarity measure used allows the first of these events to count as a positive increment to the degree of similarity of the two words and the second a decrement. In order to weight all the subjects equally, however, it is necessary to normalize the proposed increases and decreases using the mean and sum of squares of these for each subject; in this way the final similarity measure for words k and l, Z_{kl}, is given by

$$Z_{kl} = \sum_{i=1}^{N} \left(\frac{X_{ikl} - E_i}{S_i} \right) \qquad (2)$$

where N is the number of subjects and X_{ikl} is given by

$$X_{ikl} = \left.\begin{array}{c} -\log_2 P_{ij} \\[2mm] \log_2 Q_i \end{array}\right\} \text{ where words } k \text{ and } l \left.\begin{array}{c} \text{are placed in} \\ \text{the same group, } j \\ \text{are placed in} \\ \text{different groups} \end{array}\right\} \text{by subject } i$$

and E_i and S_i are the mean and sum of squares terms;

$$E_i = -\left(\sum_{j=1}^{N_i} P_{ij} \log_2 P_{ij} \right) + Q_i \log_2 Q_i \qquad (3)$$

$$S_i = \sum_{j=1}^{N_i} P_{ij}(\log_2 P_{ij})^2 + Q_i(\log_2 Q_i)^2 \qquad (4)$$

Table 1
Similarities of pairs of adjectives describing pain

	1	2	3	4	5	6	7	8	9	10	11	12	13	14	15	16	17	18	19	20	21	22	23	24	25	26	27
2	106.36																										
3	27.72	30.33																									
4	19.45	16.23	144.52																								
5	27.04	13.21	102.95	113.04																							
6	25.06	21.85	130.16	141.26	118.70																						
7	96.13	80.49	31.05	23.00	23.60	23.04																					
8	72.58	49.70	23.71	16.14	17.08	17.64	73.81																				
9	40.17	28.41	16.51	9.39	14.57	12.41	60.16	70.63																			
10	28.03	17.77	-3.61	-5.42	-4.98	-6.94	8.98	9.14	3.27																		
11	-11.40	-5.44	-11.40	-10.10	-10.10	-10.10	-5.88	-11.40	7.50	14.47																	
12	-9.67	-5.48	-1.22	-4.58	-4.58	1.07	-4.46	-2.87	26.46	20.73	100.89																
13	-3.26	-1.64	3.96	17.63	10.36	11.51	-0.74	2.85	40.82	44.64	23.44	47.08															
14	-5.44	-7.57	-5.69	-8.29	-4.36	-6.14	-2.14	-3.36	3.44	22.37	15.38	21.54	45.75														
15	-10.29	-10.29	-11.40	-11.40	-6.67	-10.25	-2.49	3.71	10.64	31.03	-4.37	5.84	50.46	39.87													
16	-9.82	-8.95	-9.82	-11.40	-9.43	-10.25	-6.87	-4.71	7.00	2.63	0.57	15.51	34.72	35.16	65.25												
17	-11.40	-11.40	-9.40	-9.40	-5.47	-8.25	-5.85	-3.32	6.89	7.82	-2.47	6.79	27.84	43.73	34.05	118.77											
18	-0.96	-2.17	-7.29	-6.95	-2.10	-6.73	0.98	1.34	-5.42	15.97	-6.99	-2.11	-5.60	-2.67	9.03	4.76	-0.98										
19	-9.27	-3.95	-2.69	-4.14	4.50	0.38	-9.09	-7.30	-4.70	-6.75	-3.97	6.41	-4.48	-2.34	-4.16	-7.30	-4.44	33.09									
20	-7.25	-1.49	-2.67	-0.74	-1.18	-3.30	-6.98	-1.96	-5.40	-5.36	11.01	5.29	-7.14	-7.98	-7.37	-8.62	-1.42	-2.37									
21	-2.80	-5.72	-4.02	-3.59	6.92	-1.21	-1.12	-4.82	-4.06	-7.76	-1.48	-2.97	-7.44	-7.07	-4.81	-6.28	-8.12	22.35	21.76	13.61							
22	-9.24	-8.23	-4.19	-6.98	-0.22	1.00	-7.48	-8.33	-3.95	-11.40	-4.72	3.86	-6.61	-6.99	-7.48	-8.56	-2.96	23.24	117.46	-8.55	11.56						
23	-5.91	-9.72	-0.18	-1.67	1.30	0.68	0.15	-11.40	-6.89	-7.92	-5.56	0.73	9.53	-7.03	-9.60	-2.96	-5.16	11.90	-1.39	6.88	6.01	2.35					
24	-11.40	-10.26	-3.75	0.40	-1.66	-2.90	-3.42	-7.30	-8.73	-10.07	-9.72	-1.29	-8.44	-7.71	-8.55	-5.91	-7.33	16.25	5.45	13.92	10.18	1.19	135.22				
25	-9.87	-8.74	-6.98	-6.71	-6.52	-8.23	-8.50	-6.76	-2.39	-9.87	-1.76	7.97	-9.37	-5.64	-9.20	-0.02	7.25	10.76	6.66	-3.89	36.50	11.92	65.80	63.89			
26	-9.94	-8.34	-9.93	10.10	-9.31	-10.10	-9.93	-6.91	-11.40	-8.50	-10.10	-11.40	-11.40	11.40	-7.41	-11.40	-9.96	-9.15	-8.35	-11.40	-11.40	-11.40	-11.40	11.40	-11.40		
27	-9.67	-9.67	-11.40	11.40	-10.61	-11.40	-11.40	-6.97	-6.94	-10.01	-11.40	-9.67	-4.47	11.40	-6.12	-10.06	-8.67	-9.21	-10.10	-11.40	-7.01	-9.67	-8.71	10.10	-8.71	162.96	
28	-9.67	-8.07	-11.40	11.40	-10.61	-11.40	-8.31	-6.85	-3.85	-10.20	-11.40	-8.48	-4.24	10.20	-4.80	-7.35	-5.96	-9.41	-8.28	-11.40	-11.40	-9.67	-10.00	-9.89	-10.00	170.71	178.41

Table 1.
(Continued)

29	-9.02	-9.02	-8.18	-6.88	-4.12	-5.73	-5.25	2.33	13.99	-2.37
	-4.86	3.25	7.53	7.72	6.09	25.14	41.90	-9.21	-10.03	8.76
	-11.40	-7.21	-6.26	-4.86	9.70	29.66	39.74	38.03		
30	23.12	25.19	-2.31	-5.58	-3.62	-5.58	6.90	17.71	2.81	54.48
	-5.76	-7.09	-1.31	-4.20	0.21	-7.19	-4.11	7.33	-8.52	-4.96
	-2.56	-9.96	-10.07	-10.07	-11.40	-4.35	-4.22	-7.51	-7.06	

1 Flickering	11 Boring	21 Cramping
2 Quivering	12 Drilling	22 Crushing
3 Pulsing	13 Stabbing	23 Tugging
4 Throbbing	14 Lancinating	24 Pulling
5 Beating	15 Sharp	25 Wrenching
6 Pounding	16 Cutting	26 Hot
7 Jumping	17 Lacerating	27 Burning
8 Flashing	18 Pinching	28 Scalding
9 Shooting	19 Pressing	29 Searing
10 Pricking	20 Gnawing	30 Tingling

The rationale behind this measure is that when a subject places two words in a very small group the event has high information and there is, consequently, a relatively large increment to the appropriate similarity; on the other hand, when two words are placed together in a fairly large group, then there will be only a small increase in the similarity if Q_i is small (i.e. there is a low probability of a given two words being in different groups) and a smaller decrease if Q_i is large. In this way the measure compensates for differences among subjects in the sizes of the groups produced.

Table 1 shows the values of this similarity index amongst all pairs from 30 pain adjectives.

Q1 Which method of multidimensional scaling should be used to analyse the matrix of similarities between adjectives describing pain?

What is sought from a multidimensional scaling solution is a set of coordinates in a particular number of dimensions and an associated distance measure which together represent as closely as possible the structure present in the observed similarity or dissimilarity matrix. In general terms this simply means that the larger the dissimilarity between two stimuli (or the smaller the similarity), the further apart should be the points representing these stimuli in the geometrical model.

To measure explicitly how well a particular set of coordinates fits a set of observed proximities, some objective function is defined; the simplest would be of the form

$$S = \sum_{i<j} (d_{ij} - \delta_{ij})^2 \qquad (5)$$

where δ_{ij} is the observed dissimilarity between stimuli i and j, and d_{ij} is the distance between the two points representing these stimuli in the geometrical model. The most usual distance measure used is Euclidean, so that

$$d_{ij} = \left[\sum_{k=1}^{d} (x_{ik} - x_{jk})^2 \right]^{1/2} \qquad (6)$$

where x_{ik}, x_{jk}, $k = 1, \ldots, d$, are the coordinates of the points representing stimuli i and j, and d is the dimensionality of the geometrical model.

Since d_{ij} is a function of the coordinate values, so too is S; minimization of S with respect to these coordinate values (using one or other numerical optimization algorithms) will lead to a model which represents the observed dissimilarities. Practical methods of multidimensional scaling do not use S directly as an objective function; instead S is scaled in some way to make it invariant under uniform stretching or shrinking of a set of coordinates (see Everitt and Dunn, 1983, for details). The scaled objective function is then often known as *stress*.

A further complication is that the observed dissimilarities are almost never used directly in objective functions such as S. If they were it would imply that it is reasonable to assume the following relationship between observed dissimilarities and fitted distances:

$$d_{ij} = \delta_{ij} + \varepsilon_{ij} \qquad (7)$$

where the ε_{ij} represent a combination of errors of measurement and distortion errors arising because the dissimilarities may not exactly correspond to a configuration in d dimensions.

Such a simple relationship may be unrealistic in many situations and in general a relationship between distances and dissimilarities of the form

$$d_{ij} = f(\delta_{ij}) + \varepsilon_{ij} \qquad (8)$$

might be postulated. Minimization of stress would now essentially involve a two-stage procedure. First we would need to minimize with respect to the parameters of f and then with respect to the coordinates. When f is a linear function, for example

$$f(\delta_{ij}) = a + b\delta_{ij} \qquad (9)$$

or

$$f(\delta_{ij}) = a + b\delta_{ij} + c\delta_{ij}^2 \qquad (10)$$

the first-stage minimization may be achieved by simple linear the first-stage minimization may be achieved by simple linear regression techniques. In such cases the scaling technique is referred to as *metric*.

A very important method of scaling, which arises by allowing f in (8) to indicate a *monotonic relationship* between distance and dissimilarity, was first suggested by Shepard (1962) and Kruskal (1964). Here the distances are related to the dissimilarities by the relationship

$$d_{ij} = \hat{d}_{ij} + \varepsilon_{ij} \tag{11}$$

where the $\hat{d}_{ij}$'s are a set of numbers monotonic with the δ_{ij}'s. This method of *non-metric* scaling has the important property of leading to solutions invariant under monotonic transformations of the dissimilarities. This is very useful for measures arising from experiments involving human subjects who can usually only be relied upon to give judgements which are ordinal.

For values of d, the number of dimensions, from 2 to 5, compare the solutions obtained by applying metric and non-metric multidimensional scaling methods to the similarity matrix of the pain adjectives.

An interesting method of multidimensional scaling suggested by Ramsay (1977, 1982) is based upon maximum likelihood methods, and enables estimates of parameter variation to be made. Additionally, formal tests of whether a particular configuration provides an adequate fit to a set of observed proximities are provided. Figure 1 shows the two-dimensional solution given by Ramsay's method for the pain adjective data.

Compare Figure 1 with the corresponding two-dimensional solutions obtained from metric and non-metric scaling, and suggest ways in which the 'closeness' of the various solutions could be assessed.

Q2   How can an appropriate value of d be chosen?

A complete solution from any application of multidimensional scaling consists of a set of coordinate values in a particular number of dimensions. Some thought therefore needs to be given to choosing the appropriate value of d. Some work has been done on producing an objective, statistical method for determining the true dimensionality, for example Spence (1972) and Spence and Graef (1974). The results, however, are not particularly satisfactory. A simple method which is often suggested is to plot stress against number of dimensions and to look for an 'elbow' in the curve, the position of this being indicative of the correct number of dimensions.

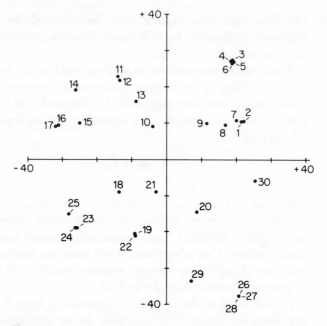

Figure 1
Two-dimensional solution from MULTISCAL for pain-adjective
data

In many respects the decision about the number of coordinates needed for a given data set is as much a substantive question as a statistical one. Even if a reasonable statistical method existed for determining the 'correct' or 'true' dimensionality, this would not necessarily in itself be sufficient to indicate how many coordinates are needed. Since multidimensional scaling is almost always used as a descriptive model for representing and hopefully understanding a data set, other considerations enter into decisions about dimensionality. This point is made by Gnanadesikan and Wilk (1969):

> Interpretability and simplicity are important in data analysis and any
> rigid inference of optimal dimensionality in the light of the observed
> values of a numerical index of goodness-of-fit may not be productive.

In the light of such a comment, two-dimensional solutions are likely to be those of most *practical* importance since they have the virtue of simplicity, are often readily assimilated by the investigator and may, in many cases, provide an easily understood basis for discussion of observed proximity matrices.

Examine the metric and non-metric solutions found for the pain adjectives similarity matrix, in two, three and four dimensions. Is there any evidence that the two-dimensional solution is inadequate?

Q3 Has stress really been minimized?

The core of most methods of multidimensional scaling involves the optimization of stress or some other objective function by some type of numerical optimization algorithm such as steepest descent, Newton–Raphson, etc. Such techniques seek the minimum of the objective function by a series of steps taken from some initial set of

Table 2
Two-dimensional coordinates arising from the MULTISCAL method applied to the similarity of pain adjectives data

ADJECTIVE	COORDINATE 1	COORDINATE 2
1	11	21
2	10	22
3	27	19
4	27	19
5	27	19
6	27	19
7	11	20
8	10	17
9	10	12
10	9	-4
11	23	-14
12	22	-14
13	16	-9
14	19	-26
15	10	-25
16	9	-31
17	9	-32
18	-9	-13
19	-21	-9
20	-15	9
21	-9	-3
22	-21	-9
23	-19	-26
24	-19	-26
25	-15	-28
26	-37	20
27	-37	20
28	-37	20
29	-34	7
30	-6	25

coordinate values either generated by some automatic process or supplied by the user. One problem that can arise is that different initial configurations may lead to different final solutions, some of which correspond to a *local*, rather than the *global*, minimum of the objective function.

Investigate this problem for the pain adjectives data by comparing solutions and stress values obtained from the initial configuration generated internally by the package you are using and that given in Table 2, which arises from the MULTISCAL method.

OTHER POSSIBLE ANALYSES

It can often be beneficial to combine the two-dimensional solution obtained from one or other method of multidimensional scaling with the results of a cluster analysis applied to the same similarity or dissimilarity matrix. A simple way to do this is to indicate the clusters on the two-dimensional diagram by drawing in appropriate boundaries. The resulting diagram is often helpful in indicating any distortions that have resulted from representing the observed relationships between stimuli in two dimensions; for example, points which are widely separated on the diagram but which are placed in the same group by the cluster analysis method.

Q4   After applying one of the available hierarchical clustering techniques to the pain adjectives similarity matrix, display the results on one of the two-dimensional diagrams obtained from one or more of the various multi-dimensional scaling analyses performed.

REFERENCES

Burton, M. (1972). Semantic dimensions of occupation names. In *Multidimensional Scaling, Volume II: Applications*, A. K. Romney, R. N. Shepard and S. B. Nerlove (eds.), Seminar Press, New York.

Carroll, J. D., and Wish, M. (1974). Applications of individual differences scaling to studies of human perception and judgement. In *Handbook of Perception, Volume 2*, E. C. Carterette and M. P. Friedman (eds.), Academic Press, New York.

Chatfield, C., and Collins, A. J. (1980). *Introduction to Multivariate Analysis*, Chapman and Hall, London.

Ekman, G. (1954). Dimensions of colour vision, *Journal of Psychology*, **38**, 467–74.

Everitt, B. S., and Dunn, G. (1983). *Advanced Methods of Data Exploration and Modelling*, Gower Press, London.

Gnanadesikan, R., and Wilk, M. B. (1969). Data analytic methods in multivariate statistical analysis. In P. R. Krishnaiah (ed.), *Multivariate Analysis, Volume 2*, Academic Press, New York.

Hair, J. F., Anderson, R. W., and Tatham, R. L. (1987). *Multivariate Data Analysis*, Macmillan, New York.

Kruskal, J. B. (1964). Multidimensional scaling by optimizing goodness of fit to a nonmetric hypothesis, *Psychometrika*, **29**, 1–27.

Kruskal, J. B., and Wish, M. (1978). *Multidimensional Scaling*, Sage Publications, Beverly Hills, California.

Levelt, W. J. M., Van de Geer, J. P., and Plomp, R. (1966). Triadic comparisons of musical intervals, *British Journal of Mathematical and Statistical Psychology*, **19**, 163–79.

Mardia, K. V., Kent, J. T., and Bibby, J. M. (1979). *Multivariate Analysis*, Academic Press, London.

Ramsay, J. O. (1977). Maximum likelihood estimation in multidimensional scaling, *Psychometrika*, **42**, 241–66.

Ramsay, J. O. (1982). Some statistical approaches to multidimensional scaling data (with discussion), *Journal of the Royal Statistical Society (Series A)*, **145**, 285–312.

Romney, A. K., Shepard, R. N., and Nerlove, S. B. (1972). *Multidimensional Scaling, Volume II: Applications*, Seminar Press, New York.

Shepard, R. N. (1962). The analysis of proximities: multidimensional scaling with an unknown distance function. I, *Psychometrika*, **27**, 125–40.

Shepard, R. N. (1974). Representation of structure in similarity data. Problems and prospects, *Psychometrika*, **39**, 373–421.

Siegel, S. (1956). *Nonparametric Statistics*, McGraw-Hill, New York.

Spence, I. (1972). An aid to the estimation of dimensionality in nonmetric multidimensional scaling, University of Western Ontario Research Bulletin 229.

Spence, I., and Graef, J. (1974). The determination of the underlying dimensionality of an empirically obtained matrix of proximities, *Multivariate Behavioral Research*, **9**, 331–41.

Wish, M. (1971). Individual differences in perception and preferences among nations. In C. W. King and D. Tigert (eds.), *Attitude Research Reaches New Heights*, American Marketing Association, Chicago.

Wish, M., Deutsch, M., and Biener, L. (1972). Differences in perceived similarity of nations. In A. K. Romney, R. N. Shepard and S. B. Nerlove (eds.), *Multidimensional Scaling, Volume II: Applications*, Seminar Press, New York.

SUPPLEMENTARY QUESTIONS

1. What is meant by the horseshoe effect in multidimensional scaling? (See Kendall, 1975.)

2. Using the following similarity matrix (from Woese, 1981), apply metric

multidimensional scaling (using a linear transformation of the data) and non-metric multidimensional scaling to produce configurations of the sixteen points in two, three and four dimensions.

```
0.29
0.33  0.36
0.05  0.10  0.06
0.06  0.05  0.06  0.24
0.08  0.06  0.07  0.25  0.22
0.09  0.10  0.07  0.28  0.22  0.34
0.11  0.09  0.09  0.26  0.20  0.26  0.23
0.08  0.11  0.06  0.21  0.19  0.20  0.21  0.31
0.11  0.10  0.10  0.11  0.06  0.11  0.12  0.11  0.14
0.11  0.10  0.10  0.12  0.07  0.13  0.12  0.11  0.12  0.51
0.08  0.13  0.09  0.07  0.06  0.06  0.09  0.10  0.10  0.25  0.25
0.08  0.07  0.07  0.12  0.09  0.12  0.10  0.10  0.12  0.30  0.24  0.32
0.10  0.09  0.11  0.07  0.07  0.10  0.10  0.13  0.12  0.34  0.31  0.29  0.28
0.07  0.07  0.06  0.07  0.05  0.07  0.06  0.10  0.06  0.17  0.15  0.13  0.16  0.19
0.08  0.09  0.07  0.09  0.07  0.09  0.09  0.10  0.07  0.19  0.20  0.21  0.23  0.23  0.13
```

(This illustrates the problem of *degeneracy*; see Kruskal and Wish, 1978.)

3. How might the definition of an objective function such as stress be altered to accommodate (a) missing values in the proximity matrix and (b) ties in the proximity matrix?

4. In an ordinal MDS analysis of six stimuli, the following figures are the output from the isotonic regression, generated by a two-dimensional solution. Construct the Shepard diagram from them, clearly labelling all relevant features. What interpretations can be made from the diagram and what further information would be required to judge adequacy of this solution?

STIMULI	AB	AC	AD	AE	AF	BC	BD
Dissimilarity	10.0	17.7	36.5	33.0	52.0	10.9	18.2
Disparity	5.5	9.4	17.7	17.7	25.4	7.1	11.8
Distance	5.5	8.4	17.2	17.9	25.4	7.1	11.8

STIMULI	BE	BF	CD	CE	CF	DE	DF	EF
Dissimilarity	28.4	45.5	20.6	14.0	40.9	48.5	47.2	16.4
Disparity	16.5	20.8	16.5	9.4	17.7	20.8	20.8	9.4
Distance	16.5	22.0	16.5	9.7	17.9	22.0	22.4	10.1

5. A method suggested by Sammon (1969) for obtaining a low-dimensional representation of multivariate data is to seek a set of

coordinates $x_1, \ldots, x_{p^*}$ for each individual so as to minimize

$$S = \sum (d_{ij} - d_{ij}^*)^2$$

where

$$d_{ij} = \left[\sum_{k=1}^{p} (y_{ik} - y_{jk})^2 \right]^{1/2}$$

and

$$d_{ij}^* = \left[\sum_{k=1}^{p^*} (y_{ik} - y_{jk})^2 \right]^{1/2}$$

are the Euclidean distances between pairs of points for the original data and for the derived coordinates. Construct a computer program for implementing this procedure and use it to find a two-dimensional representation of the five-dimensional observations given in the table below:

	1	2	3	4	5
1	1.79052	2.40779	2.30405	3.33947	4.81159
2	0.41870	2.41622	−1.15054	1.60016	3.77954
3	0.38100	2.17836	−1.51034	1.50106	5.23665
4	1.40827	2.54015	−0.82001	1.21235	2.22920
5	0.98923	2.23567	−1.46415	2.25915	1.72468
6	0.72403	2.12971	−1.32712	2.56672	1.66196
7	1.93870	2.27699	−0.63780	2.62517	2.91837
8	−0.54793	1.61704	−1.34134	1.93518	3.66258
9	0.74240	2.01851	−2.35586	1.28895	3.80879
10	1.76550	2.40141	−0.47575	3.03373	1.96689
11	1.16110	−0.91696	1.13762	−0.65443	0.93723
12	2.91133	−0.79694	2.68309	0.73394	1.44418
13	2.50516	0.46141	0.37595	0.06101	0.79193
14	1.86969	1.05876	1.14048	−0.50032	0.74940
15	1.82928	−0.05657	1.21313	0.12843	0.89604
16	2.26915	0.22230	1.84513	−0.15286	0.62941
17	2.06271	−0.64234	1.59496	0.52112	1.66441
18	1.29071	0.84857	1.37032	0.65502	1.46672
19	1.60536	0.84620	2.92455	0.15850	1.00242
20	1.60489	0.10619	1.78494	−0.41721	0.99682

FURTHER REFERENCES

Kendall, D. G. (1975). The recovery of structure from fragmentary information, *Philosophical Transactions of the Royal Society (Series A)*, **279**, 547–82.

Sammon, J. W. (1969). A nonlinear mapping for data structure analysis, *IEEE Transactions on Computers*, C-18, 401–9.

Woese, C. R. (1981). Archaebacteria, *Scientific American*, **244**(6), 94–106.

THE DESIGN AND ANALYSIS OF SURVEYS

Assignments in Applied Statistics
Edited by S. Conrad
© 1989 John Wiley & Sons Ltd

Introduction

Stephanie Stray

School of Industrial and Business Studies, University of Warwick

The results of surveys increasingly impinge on one's everyday life. Newspapers and television use survey results in a wide variety of ways and the advent of relatively cheap computing power, together with an increasing variety of statistical software, has meant that the analysis of survey data can be effected much more quickly and undertaken by a wider variety of individuals and institutions than has been true in the past. Almost everyone will have seen the results of some poll either on television or in a newspaper and many people will have had market researchers, or researchers from some specific organization or group, knock on their door or accost them in the street with the aim of eliciting information. The old maxim of 'familiarity breeds contempt' is probably aptly applied to this subject since there appears to be an increasing suspicion of survey results. In many ways this suspicion is well founded since badly designed surveys with inappropriate sample designs, poor questionnaires and shoddy analysis cannot be expected to aid effective decision making.

Despite the commonplace nature of surveys in contemporary society, it is only during the course of this century that they have been conducted in the sort of format we would typically recognize today. The concept of a *respondent*, where individuals are asked to provide information about themselves, is one that was, in this country, firmly established only relatively recently. Consequently, unlike the United States where survey research had a somewhat earlier start, it was not until after the Second World War that in Britain any major developments came in the data collection aspects of the survey process. For an excellent review of the history of the use of surveys in Britain see Marsh (1982), pp. 9–47. The prime impetus came from what is now known as the Office of Population Censuses and Surveys (OPCS) (formed in 1970 from the amalgamation of the Social Survey—which was originally a part of the Central Office of Information—and the Registrar General's office) and from the growth of the market and consumer research (and opinion polling) industry.

The purpose of a survey can be either descriptive or explanatory or both. A descriptive survey is essentially a fact-seeking exercise and is designed to answer such questions as how much people spend on food, what dog foods people buy, how different socioeconomic groups spend their incomes, etc. An explanatory survey may be designed to test out some hypothesis generated by a specific sociological or economic theory or designed to examine causal relationships between variables such as the influence of stress on heart disease or the effects of changes in product packaging upon sales. In either case, actually conducting a survey presents one with a minefield of problems and decisions. What are the clients needs? What data need to be collected, from whom and how? How much data need to be collected? Who or what should be used to collect the data? What resources are available? How much time is available to conduct the fieldwork and process the results? How should the data be analysed? These are just a few of the myriad of questions and details that need to be decided, along with many others, in the course of the survey process.

Regardless of whether one is conducting a survey that is descriptive or explanatory (or both), there is a need to collect data upon *elements*. An element is the unit upon which information or data is sought. The *population* is the collection of elements about which one wishes to make a statement or series of statements. There may be a need to distinguish between the *theoretical population* that one wishes to study and the *survey population* which refers to the collection of elements that one actually studies in practice. The major differences are likely to be due to non-response and non-coverage. Therefore certain age groups amongst the population may be omitted as being too young or too old to supply the necessary information, or people living in certain geographical areas may be excluded due to problems of accessibility. One also needs to distinguish between *elements* and *sampling units*. Sampling units are used to select elements into the sample. In some instances these may be the same as the elements and in other instances the two may differ. Therefore one may select a sample of addresses and then collect information from all individuals living at those addresses. If individuals are the basic unit about which information is sought, then the addresses are the sampling units and the elements are the individuals at those addresses. Therefore sampling units are collections of elements in the population with each element belonging to one and only one sampling unit. A sampling *frame* is a list of all the sampling units. Common examples of sampling frames that are used are the electoral register, rating lists and telephone directories. All of these invariably have shortcomings in that they rarely provide a complete list of the desired sampling units or may include entries that are not in the population of interest. A *sample* is a collection of sampling units that is selected from the frame. More strictly, this is the procedure that is typically undertaken in the case of probability sampling, where each element has a non-zero chance of being selected into the sample. For non-probability sampling methods (quota sampling, fortuitous sampling and judgemental sampling for instance), samples are selected by other means.

Probability samples may be selected in a variety of ways. The basic sample design is that of *simple random sampling*. This form of sampling may take place either with or without replacement but underpins the methods of statistical inference that are covered in most courses on statistics. However, it is rarely used in its simple unadulterated form in survey design.

Stratified sampling is one of the most commonly employed procedures in survey design. In this case the population—or frame—is subdivided into distinct subgroups or strata and samples are selected from each of the strata. Efficient use of this approach requires that the population is divided into groups which are different from each other with respect to the variable, or variables, of interest.

Cluster, or *multistage, sampling* is also commonly employed, especially in surveys where the population is geographically dispersed. In this form of sampling the population is again divided into separate groups as in the case of stratified sampling. These groups are known as the *first-stage units* and the subdivision of the population into these first-stage units frequently takes place on the basis of the geographical location of the sampling units. The first stage of the selection process involves selecting only a limited number of the first-stage units. In the case of a *two-stage* sample design sampling units are selected from the selected first-stage units. Thus the sample is composed only of sampling units in *some* first-stage units, not in all first-stage units. In the case of a *three-stage* sample design each of the selected first-stage units are further subdivided into groups—known as the *second-stage units*—and a limited number of these second-stage units are chosen from each of the selected first-stage units. The selection of sampling units from the chosen second-stage units then takes place and forms the third stage in the selection process. Multistage sampling can have as many stages as one desires but between two and four stages is the most common. The essential difference between multistage and cluster sampling is that in the case of cluster sampling the final stage consists of *complete enumeration* of the units selected at the penultimate stage. Therefore, in the case of a two-stage cluster sample *all* the sampling units are selected from each of the selected first-stage units, rather than a sample of units. The primary advantage of multistage or cluster sampling is that the resulting sample is confined to a limited number of geographical areas and therefore the data collection process may be considerably cheaper than if the sample is widely dispersed over a large geographical area.

A final type of probability sampling that is considered in the assignments is that of *systematic* sampling. Whilst there are many variations in this type of sample selection method the basic means of selecting the sample in this type of design is similar throughout. Essentially a particular point on the sampling frame is located and then the sampling unit at that point is selected together with every kth subsequent sampling unit. Thus every tenth, twentieth or hundredth item is chosen until sufficient sampling units have been selected.

The one example of a non-probability sample design that is referred to in the assignments here is that of *quota* sampling. In quota sampling various background characteristics of the population are chosen—such as age, sex, etc.—and it is ensured that the sample reflects the distribution of people in the population with these characteristics. Thus if 54 per cent of the population are female, then 54 per cent of the sample consists of females. Similarly, if 30 per cent of the population is between 18 and 35 years of age, then the same proportion of the sample lies in this age group. However, there is no element of random sampling in the selection procedure. Interviewers are given a quota informing them how many people of different types they must interview—the number of males, the number of females, etc.—and are free to select whoever they wish in order to satisfy the quota they have been given. Thus the chance of selection for a particular individual is not known in advance but depends upon where the interviewer is located and upon the personal biases and preferences of the interviewer. The primary advantage of this form of sampling is that no sampling frame is needed and that the data collection process is relatively quick since no interviewer call-backs are necessary. If someone refuses to cooperate or is not at home at the time an interviewer calls then the interviewer can substitute another individual who fits the background characteristics that have been given.

For further details of all the types of sample design described above, including discussions of the relative merits of the designs, see Moser and Kalton (1971). For a more statistical treatment including details of the standard errors of the above designs (where appropriate) see either Cochran (1963) or Kish (1965).

Regardless of the type of sample design that is employed, surveys invariably require the use of a questionnaire or of an interviewing schedule. Questionnaires may be distributed through the post or individuals may be contacted by an interviewer and asked to participate in the survey. The design of the questionnaire or interviewing schedule is an exceedingly important part of the overall survey process since it is via this mechanism that one gains the necessary data for analysis. Poor questionnaire design is likely to lead to useless data. Furthermore, interviewing schedules and self-completion questionnaires may need to be designed differently since the form and type of question one can ask using the former can be very different to the latter. For further details of questionnaire design see Oppenheim (1966).

The four assignments in this section cover various aspects of the design and analysis of surveys. The first assignment concerns the British Election Study of 1979. The Election Studies were a series of surveys that were started in the 1960s and continued on through the 1970s and 1980s. Each survey took place immediately after a general election in Britain and was designed to provide data to enable the major factors affecting voting decisions by the British electorate to be determined and investigated. The Studies were also designed to enable researchers to determine the socioeconomic basis of each political party's support. The Studies of the 1970s, which took place following each of the three general elections of February 1974, October 1974 and May 1979, also had a second aspect associated with them. Respondents who were interviewed following the first of these three elections were contacted, where possible, for the Election Studies that took place following each of the next two elections. Therefore, apart from having data collected from a cross-section of the electorate at the time of each election, data were also available from the same individuals at different points in time. This enables changes in voting behaviour, opinions and attitudes to be monitored and a much more detailed understanding of British psephology to be gained.

The assignment describes the survey design that was used in February 1974 and also explains how this was modified for the survey that took place in 1979. After a lapse of more than five years many of the original respondents proved difficult to trace—such panel attrition being a common problem of panel research—and so the original sample of people needed to be supplemented to maintain the sample size and to ensure the representativeness of the overall sample.

As has already been mentioned above, the media often use polling organizations to provide them with information not just upon voting intentions of the British electorate but also about public opinion on topical subjects and matters of current concern. The second assignment concerns opinion poll surveys and describes two types of samples that are used by National Opinion Polls (NOP) and gives an example of a typical questionnaire used by such polling organizations. Part of the intention of the assignment, apart from illustrating other types of sample designs, is to focus attention on the interpretation that one can place on the results of such polls.

Survey methods and survey research have obviously long been bound up with using and exploring the traditional means of data collection using face-to-face interviewing and mailed questionnaires. More recently telephone sampling has become increasingly common in Britain, despite the criticisms frequently levelled at this form of survey sampling. Furthermore, the increasingly widespread use of microcomputers and the declining cost of computer power are leading to changes in the traditional approaches to data collection. The third assignment in this section is concerned with the traditional survey process and the changes that the use of microcomputers are bringing to bear upon this traditional process.

The final assignment is concerned with analysing survey data. There is no value in collecting data if one has little or no idea of what to do with it once it is available. Much of the data obtained from surveys by market researchers, opinion pollsters, government organizations, individual companies and private researchers is categorical in nature. Hence the focus of this fourth assignment is on data of this type and on the analysis of such data. No sophisticated statistical knowledge is assumed although some of the pitfalls and problems of analysis are discussed and illustrated. The assignment also introduces log-linear models and the basic concepts employed by such models.

Thanks are due to the NOP for their kind permission to reproduce some of their material, to Professor Ivor Crewe and Professor Bo Sarlvik for their permission to use information from the British Election Studies, as well as to the Data Archive at the University of Essex, and to Professor Wilem Saris for permission to use information collected by the Sociometric Research Foundation.

REFERENCES

Cochran, W. G. (1963). *Sampling Techniques*, Wiley, New York.

Kish, L. (1965). *Survey Sampling*, Wiley, New York.

Marsh, C. (1982). *The Survey Method*, Allen and Unwin, London.

Moser, C. A., and Kalton, G. (1971). *Survey Methods in Social Investigation*, Heinemann, London.

Oppenheim, A. N. (1966). *Questionnaire Design and Attitude Measurement*, Heinemann, London.

Assignments in Applied Statistics
Edited by S. Conrad
© 1989 John Wiley & Sons Ltd

Sample Design and the 1979 British Election Study

Stephanie Stray

School of Industrial and Business Studies, University of Warwick

OUTLINE

The series of British Election Studies provides vital information for the study of voting behaviour in Britain and is one of the major sources of data for electoral research. This assignment concerns the 1979 Election Study and focuses on the sample design used. The 1979 Study also formed part of a series of Election Studies that interviewed a panel of people at successive elections in order to monitor individuals' behaviour and attitudes over time. Hence the sample taken in 1979 had to be representative of the 1979 British electorate as well as providing information on the panel element. Details of the sample design used in the Election Study of February 1974 are also given since this sample comprised the beginning of the panel study. This enables the modifications to the sample selection process in 1979 to be evaluated in the context of the needs to continue the panel study and to provide accurate and reliable information on the 1979 election.

KEYWORDS

British Election Study, panel data, interviewing, sampling fraction, non-response, stratification, multistage sampling, self-weighting, reweighting, sampling frame

The 1979 British Election Study was one of a series of surveys that started in the 1960s and continued through the 1970s. Each survey took place at the time of a General Election in Britain with the aim of studying the voting characteristics of the British electorate (i.e. excluding Northern Ireland) and monitoring changes in voting behaviour over time. Beginning with the Study of the February 1974 General Election and continuing with the Studies of the October 1974 Election and the May 1979 Election a panel element was introduced into the surveys. This meant that some of the same individuals were surveyed on each of these three successive occasions in order to monitor the changes in attitudes, opinions and voting behaviour of different types of respondents.

Part of the aim of the 1979 Study was to continue to ask many of the questions that had previously been put to respondents in the 1974 Studies. The responses would then enable any aggregate changes in voting behaviour, attitudes and opinions about the political parties and changes in the importance of particular issues to be determined. However, changes in aggregate figures are almost certain to disguise much of the volatility that takes place amongst the electorate.

For instance, consider a situation where aggregate figures from two successive surveys showed a rise of three percentage points in the percentage of people supporting one particular political party. This does not necessarily mean that all people previously supporting that party at the first election also voted for it on the second occasion together with an additional 3 per cent of people previously supporting other parties (or who did not vote at all). It is much more likely that many more than 3 per cent of the people changed their minds about which party to support. Some people who had supported the party in question the first time round would, on the second occasion, cast a vote elsewhere (or abstain), whereas others would be attracted to the party in question despite previously having given their support to another party (or not voted). The *net* effect is a rise of three percentage points in the support for the party under consideration, although many more than 3 per cent of the people changed their minds. It requires a panel of individuals studied over time to ascertain the more volatile nature of the underlying changes in voting behaviour since such changes cannot be determined solely from aggregate figures concerning samples of different individuals at two elections.

Whilst the 1979 Election Study was concerned with monitoring a panel of individuals and changes in their responses to particular questions, it was also concerned with issues salient to the 1979 election and hence new questions were asked that had not formed part of the questionnaire used in the previous studies of 1974. For instance, Mrs Thatcher's leadership of the Conservative Party meant that one aspect of importance in 1979, that was not germane to earlier studies, concerned whether or not the sex of the

Prime Minister was thought to be important for his or her suitability for the position. Would electors be more or less likely to support the Conservative Party because it had a female leader or did this make no difference to the way people voted? Other issues about which information was sought for the first time in 1979 concerned devolution and future economic and political developments within Britain.

In the 1979 Study information was sought not just upon voting behaviour but also upon respondents' attention to various aspects of the political campaign and their knowledge of and liking (or disliking) of the various political parties and leaders. Questions were also asked upon respondents' attitudes towards possible changes in the political system and upon issues such as taxation, prices, employment, race relations, strikes, social services, council house sales, capital punishment, nationalization, Britain's EEC policies and trade union activities. The salience of many of these issues in an individual respondent's vote choice was also sought together with information on beliefs about the policy positions of the various political parties upon these matters. Finally, details were also sought about the respondent's own sociopoliticoeconomic background characteristics such as income, education, social class, religion and past voting behaviour.

Therefore the 1979 Study had twin purposes. First, it aimed to contact as many people as possible who had previously been interviewed as part of the Election Studies of February and October 1974. These people were to form the panel element of the study. However, mortality, residential mobility and willingness to cooperate (or the lack of it) meant that not all the previously contacted individuals were available for further study.

Second, the 1979 Study also aimed to achieve a representative cross-section of the 1979 electorate. Changes in the nature and composition of the British electorate in the five years since 1974 due to the entry of a new age cohort of individuals on to the Electoral Register, and mortality and geographical mobility mean that, apart from problems of panel attrition since 1974, the respondents who had previously been interviewed would no longer comprise a representative cross-section of the electorate in 1979. Therefore there was a need to supplement the original panel of respondents with new individuals in order to ensure the representativeness of the overall sample. What follows below is a description of the survey design used in the February 1974 Election Study together with details of how this was modified in 1979 in respect of the aims and problems outlined above.

THE FEBRUARY 1974 ELECTION STUDY

In February 1974 the 618 constituencies in Great Britain south of the Caledonian Canal were grouped into eleven Standard Regions with

Greater London being treated as a separate region. Within each region constituencies were divided into three Local Authority groupings; these being conurbations, urban areas and rural areas. This division was undertaken according to the percentage of a constituency's population living in each of these three types of Local Authority Areas. The type of Local Authority Area that contained the greatest proportion of a constituency's electorate determined the classification of that constituency. This led to a total of 26 groups since there were no conurbations in five of the Standard Regions (the West Midlands, East Anglia, the South East, the South West and Wales) and because all Greater London constituencies were classified as conurbations. Within each of the 26 regional groupings constituencies were listed in descending order of the Labour vote at the 1970 election (which immediately preceded the election of February 1974). For those cases where constituencies had been affected by the revision of constituency boundaries between 1970 and 1974 the 1970 voting figures were used for the old constituency that contained the greatest number of people currently in the new constituency.

The procedure that was then adopted was to select 200 constituencies from the 26 regional groupings. The number of constituencies selected from each regional grouping was required to be in proportion to the total electorate in each regional area. Therefore more constituencies were selected from the regions containing a large proportion of Britain's electorate and fewer constituencies were selected from regions containing a small proportion of the electorate. From each of the 200 selected constituencies one polling district was chosen and from each of the resulting 200 polling districts seventeen names were selected from the polling district's Electoral Register. The methods by which the constituencies, then the polling districts and, finally, the names were selected is described below.

To select constituencies from a regional group the constituencies for a group were listed (in descending order of Labour's vote in 1970) and their 1973 electorates were cumulated. Table 1 gives an example for the first five constituencies in a particular regional grouping and shows for each constituency the 1970 Labour vote, the 1973 electorate and the cumulated 1973 electorates. The cumulated electorates were used to assign a range of values to a particular constituency—the range being equal to the number of people in that constituency's electorate. For example, as can be seen in Table 1, the first constituency, A, has an assigned range of 1 to 58,112 (there being 58,112 eligible voters in the constituency). The second constituency, B, continues with the sequence of numbers and starts at 58,113 (spanning a total of 60,151 numbers which is the same as the electorate for constituency B). The third constituency, C, starts at 118,264 and spans 57,089 numbers up to 175,352—the value of the cumulated electorate—and so on for the remaining constituencies.

Table 1
The Labour vote and electorates for the first five constituencies in one regional group

CONSTITUENCY	1970 LABOUR VOTE (%)	1973 ELECTORATE	CUMULATIVE ELECTORATE	ASSIGNED RANGE
A	73	58,112	58,112	1–58,112
B	68	60,151	118,263	58,113–118,263
C	64	57,089	175,352	118,264–175,352
D	58	55,339	230,691	175,353–230,691
E	57	58,693	289,384	230,692–289,384

The next step was to divide the total number of electors in a regional grouping by the number of constituencies to be selected from that group to give a number, f. A random number between 1 and f ($= k$) was then selected in order to generate the sequence of numbers k, $k + f$, $k + 2f$, $k + 3f$, etc. The series was continued until the number of values in the series equalled the number of constituencies to be selected from the regional grouping under consideration. Therefore, for a regional grouping with 1,600,000 electors from which eight constituencies were to be selected the value of f is 200,000. If the random number chosen (between 1 and 200,000) transpires to be 63,162 then the sequence of numbers is 63,162, 263,162, 463,162, 663,162, etc. This sequence can then be used to select constituencies from the regional group based upon the assigned ranges attached to individual constituencies, as illustrated in Table 1. Since the first number in the sequence is 63,162 and since this falls in the range assigned to constituency B, then constituency B is selected for inclusion in the sample. The second number in the sequence, 263,162, falls in the range assigned to constituency E, which becomes the second constituency to be selected for the sample.

The above method was used for each of the 26 regional groupings and ensures that the chance of an individual constituency being selected is proportional to the size of the electorate in that constituency. Constituencies with large electorates have a greater chance of being selected than constituencies with small electorates.

For the 200 constituencies that were selected in this manner a similar procedure was used in order to select a single polling district from each of these 200 constituencies. Polling districts for a particular constituency were listed in the order in which they appeared on the Electoral Register and the electorate determined for each polling district. The cumulative electorates for the polling districts were also calculated in order to assign a range of values to each polling district in a similar way to the method described above concerning the assignment of ranges of values to constituencies. A single random number between one and the number of electors in the constituency was then selected and this random number determined, on the basis of the ranges of values assigned to the polling districts, which particular polling district was to be selected.

The final task was to select individual electors from each of the 200 selected polling districts. This was done by dividing each polling district's electorate by 17 to give a sampling fraction f. A random start point on the Electoral Register for the polling district in question was then determined and that individual was selected followed by every subsequent fth individual. Once the end of the Register for a particular polling district was encountered, selection continued by returning to the start of the Register for that polling district and continuing the count for every fth individual until seventeen names (with their associated addresses) had been selected. In some instances the selection was of a 'Y-voter'. These 'Y-voters' were individuals attaining the age of 18 during the life of the Electoral Register and who were not necessarily eligible to vote in a 1974 election unless the election took place after their eighteenth birthday. In these instances the name was selected if the person had a birthday after 28 February (the date of the general election); otherwise the name was replaced.

This selection procedure yielded a list of 3400 names to be contacted. A first series of interviews secured the cooperation of 2069 individuals and a second series the participation of a further 397 of the remainder. The total of 2466 interviews thus obtained resulted in 2462 usable results. All the interviews were conducted between 8 March and 18 May 1974 following the General Election on 28 February 1974. Table 2 summarizes the major reason for non-response, the single most common reason being refusal by the respondent.

Q1 Draw your own diagram or chart to demonstrate the procedure used to select the sample for the February 1974 Election Study. Indicate the various stages of the selection process and the various subdivisions of the population that were used. Mark on the diagram any salient features of the sample selection process.

Q2 What stratification factors were used in the February 1974 sample design? What is the purpose of using such stratification factors and how useful or relevant do you believe those used in the 1974 design to be? How many stages were employed to select the sample and what were these stages? What is the purpose of such multistaging? (See Moser and Kalton, 1971, pp. 85–117.)

Q3 Show for the 1974 Study that the chance of selecting an individual elector from the population is the same for all members of the electorate regardless of the regional grouping, constituency or polling district in which an individual elector lives.

Q4 What are the relative strengths and weaknesses of the February 1974 sample design?

Q5 If you wished to use the 1974 sample data to estimate the proportion of the electorate in Britain that voted for a particular political party in February 1974 how would you determine this estimate? What factors associated with the sample design will affect the accuracy of this estimate?

Table 2
Response and reasons for non-response for the February 1974 Election Study

Total number of names	3400
Total number of interviews	2466
Total number of non-respondents	934
Reason for non-response:	
Refusal by respondent	440
No reply after three or more calls	106
Respondent moved	100
Refusal by other member at same address	81
Respondent too ill	74
Respondent deceased	43
Respondent out for three or more calls	36
Respondent away for fieldwork period	32
Severe language problems	10
Address not located, premises empty or demolished	8
Other	4
	934

THE 1979 ELECTION STUDY

For the Election Study of 1979 the sample of names drawn from the 1973 Electoral Register was no longer representative of the 1979 electorate. Additional people had obviously reached the voting age in the intervening period whilst other members of the electorate had died and, also, new housing and hence new households had come into being since 1973. Furthermore, the list of names generated in 1974 obviously contained people who had moved house since that time and whose new addresses were unknown. Attempts to trace such people were obviously likely to be expensive and involve considerable time and effort and were unlikely to have a high success rate. Therefore, in order to maintain the panel element in the Studies and to ensure that the 1979 sample formed an adequately sized sample that was a representative cross-section of the 1979 electorate, it was necessary to supplement the 1974 sample to take account of population changes, new housing and the diminution in the sample size as a result of movers. Consequently, it was decided to use the *addresses* of the 1974 sample as a basis for the 1979 Study. This resulted in the sample for the 1979 Study being composed of three components.

First, any individual who had been included in the 1974 sample and who was still living at their 1974 address was included in the sample for 1979. Second, where a member of the 1974 sample had moved house, a new respondent was selected into the sample from the household that had moved into the 1974 address. Finally, an additional sample of *addresses* was selected from the 1979 Electoral Register from the same polling districts that were used for the 1974 Studies. The addresses of the resulting selections were compared with the 1973 Register. Any address that existed as a household address in 1973 was deleted from this supplementary sample. Therefore the remainder constituted a sample of addresses of new residences that had come into existence between 1973 and 1979. One individual was selected from each of these new addresses for inclusion in the sample.

The above details concern the construction of a list of *addresses* from which the 1979 sample was to be selected. The remaining task was to select *individuals* from these addresses. This was done by compiling, for each address, a list of names of the people who were eligible to vote. For the 1974 addresses this was done on the basis of the 1973 Register and for the supplementary sample of new housing the 1979 Register was used. Each interviewer was supplied with a form for each address that listed the names of the eligible electors at that address. The names were listed according to their order on the Electoral Register. These forms also contained a number of blank lines. The interviewer updated the list of names at any one address by deleting the names of those people no longer living there and adding on the blank lines the names of electors who had moved into the address or, in the case of the 1974 addresses, had come of age in the intervening period. Obviously this meant that in some cases an original household was replaced by an entirely new household.

Asterisks on the forms denoted the person at a particular address who was to be interviewed. On the forms for the 1974 addresses the 1974 respondent was always marked with an asterisk. For the 1979 supplementary sample an asterisk was put by one name on the list for each address. Asterisks were also placed at equal intervals on three of the blank lines used to list the new electors at each address. Consequently, the 1974 respondents were interviewed wherever possible, and one pre-selected person was also interviewed, if possible, from the 1979 supplementary sample of addresses. New members of the electorate and new residents were interviewed if their name happened to be beside an asterisk. This meant that in some cases more than one person was interviewed at the same address.

As an example of the procedure consider a household with four eligible electors, one of whom has been selected for inclusion in the sample. The names of these four electors would be written on the first four rows of the list supplied to the interviewer, with one name—assume this is the second name—having an asterisk beside it to indicate the person to be interviewed. New electors and new residents at the address would be listed by the interviewer starting at row five. Assume that asterisks are also placed by rows 6, 10 and 14. Any elector in rows 1 to 4 inclusive who no longer lived at the address would be deleted. The second individual (the one originally marked by an asterisk) would be interviewed (if possible) if they still lived at the address. If there were sufficient new electors and new residents listed in row 5 onwards to yield names by rows 6, 10 or 14, then the interviewer would also attempt to secure an interview with these individuals.

The overall effect of the above procedure was that panel members were interviewed wherever possible and a panel individual who had moved was replaced by a new individual living at the same address. Those enfranchised since 1974, as well as those living in new residences, were also included in the sample. If the same interval between asterisks had been used on each form then the net result of this procedure would have been to overrepresent large households. This bias was corrected by adjusting the intervals between the asterisks on the forms used by interviewers such that the intervals were made equal to the number of electors originally listed at each address (four in the example given above). The result of this form of sample selection was to produce a self-weighting random (EPSEM) sample (see Moser and Kalton, 1971, pp. 111–16 for a similar example) where each elector had the same chance of being selected for inclusion in the sample. Further details of such types of sample design can also be found in Blyth and Marchant (1973). The consequence of this type of sample design is that all electors in each of the regional groupings have the same chance of being selected into the sample and no reweighting of the sample results is required at the analysis stage. For details of reasons why reweighting of survey results may be required and a summary of the approaches used, see Sharot (1986) and Upton (1987).

For the 1979 Election Study the above sample design yielded 1756 people to be interviewed from the original February 1974 sample of 3400 (the majority of the remainder having moved out and 232 having died). A further 1153 names were added as replacement respondents either for those having moved since 1974 or as new voters since 1974, and an additional 198 names were generated as a result of using the 1979 Register for new addresses. Therefore the total number of people to be interviewed was 3107.

Q6   What changes were made to the 1974 sample design in 1979? Why were these changes made?

Q7   Show that for the 1979 Study each elector has the same chance of being selected for inclusion in the sample.

Q8   What problems arise in using the Electoral Register as a sampling frame? How might these affect the conclusions drawn from the 1974 and 1979 data? (See Moser and Kalton, 1971, pp. 154–66.)

Table 3
Response rates for the 1979 Election Study

	NUMBER TO BE INTERVIEWED	NUMBER INTERVIEWED	RESPONSE RATE (%)
Names from 1974 sample of names	1756	1003	57
Names obtained from addresses from 1979 Register	198	139	70
Names obtained from enumerating household members	1153	751	65
Total	3107	1893	61

Table 4
Breakdown of 1979 cross-section sample data by age and sex

	SEX		
AGE	MALE	FEMALE	TOTAL
18–29	169	192	361
30–59	534	533	1067
60 and over	204	245	449
Total	907	970	1877

Number of missing observations = 16

Q9 What differences does it make to the sample designs that names were selected from the Electoral Register in the February 1974 Study and that addresses were selected from the Register in the 1979 Study?

The 1979 General Election was held on 3 May and interviewing for the Election Study commenced on 5 May. Not all the 3107 individuals were able to be interviewed despite a minimum of four calls being made to addresses where the potential respondent was not at home when the interviewer called. The major reason for failing to secure interviews was, however, due to the interviewee refusing to cooperate, and this accounted

Table 5
Breakdown of 1979 cross-section sample data by region

REGION	NUMBER OF RESPONDENTS
Scotland	147
Wales	103
North[a]	507
Midlands[b]	327
South[c]	582
Greater London	211
Total	1877

Number of missing observations = 16

[a] Yorkshire and Humberside, the North-West and the North-East.
[b] West Midlands, East Anglia and the East Midlands.
[c] South-East, excluding Greater London, and the South-West.

Table 6
Breakdown of 1979 cross-section sample data by vote and 1979 election results

VOTE AT 1979 GENERAL ELECTION	SAMPLE DATA	ELECTION RESULTS
Conservative	732	13,697,923
Labour	586	11,532,218
Liberal	215	4,313,804
Other parties	25	987,010
Did not vote	269	9,537,490
Total	1827	40,068,445

Number of missing observations = 66

Table 7
Breakdown of 1979 panel data by age and sex

	SEX		
AGE	MALE	FEMALE	TOTAL
22–29	19	17	36
30–59	271	272	543
60 and over	136	145	281
Total	426	434	860

Number of missing observations = 6

for about 60 per cent of the non-responses. However, 1893 interviews were obtained, yielding an overall response rate of 61 per cent, which is somewhat below that of the February 1974 Study. Table 3 indicates the response rates for different categories of respondents.

One of the problems associated with all surveys concerns the underlying representativeness of the sample of respondents from whom information is collected. Whilst non-response can seriously distort the resulting composition of the sample, other factors such as inadequate sample frames, interview bias and question wording can all affect the overall accuracy of the results that are obtained. However, it is important to check the resulting sample against known background details of the population from which the sample was taken. Tables 4, 5 and 6 provide details of

Table 8
Breakdown of 1979 panel data by region

REGION	NUMBER OF RESPONDENTS
Scotland	77
Wales	58
North[a]	240
Midlands[b]	143
South[c]	261
Greater London	81
Total	860

Number of missing observations = 6

[a] Yorkshire and Humberside, the North-West and the North-East.
[b] West Midlands, East Anglia and the East Midlands.
[c] South-East, excluding Greater London, and the South-West.

Table 9
Breakdown of 1979 panel data by vote

VOTE AT 1979 GENERAL ELECTION	NUMBER OF RESPONDENTS
Conservative	362
Labour	299
Liberal	98
Other parties	13
Did not vote	79
Total	851

Number of missing observations = 15

breakdowns of the respondents by age and sex, by Standard Region and by vote, together with voting results for Great Britain respectively. Tables 7, 8 and 9 provide the same information for the panel element of the sample. These last three tables include only those individuals who had previously been successfully contacted in February 1974 as well as in 1979.

Q10   What are the response rates for the 1974 and 1979 Studies? Why would you anticipate that these response rates may differ and what factors are likely to be responsible for these differences? (See Moser and Kalton, 1971, pp. 166–86.)

Q11   What can be said about the representativeness of the 1979 sample? Use appropriate data in Great Britain to support your conclusions.

Q12   What differences, if any, exist between the composition of the panel and the non-panel elements of the 1979 survey? What conclusions can you draw, if any, from your comparison?

REFERENCES

Blyth, W. G., and Marchant, L. J. (1973). A self-weighting random sampling technique, *Journal of the Market Research Society*, **15**, 157–62.
Moser, C. A., and Kalton, G. (1971). *Survey Methods in Social Investigation*, Heinemann, London.
Sharot, T. (1986). Weighting survey results, *Journal of the Market Research Society*, **28**, 269–84.
Upton, G. J. G. (1987). On the use of rim weighting, *Journal of the Market Research Society*, **29**, 363–6.

SUPPLEMENTARY QUESTIONS

1. What are the most commonly used sampling frames for national surveys in Britain? What advantages, disadvantages and problems do each of these frames have? (See Moser and Kalton, 1971, pp. 54–66; Hakim, 1982, pp. 8–16.)
2. What would be the advantages and disadvantages of using telephone interviewing for the British Election Study? (See Miller, 1987.)
3. What are the major reasons for non-response in a survey and what problems does it cause? How would you ensure that non-response is kept to a minimum? (See Bogart, 1967; Moser and Kalton, 1971, pp. 166–86 and 256–69; McDaniel, Madden and Verille, 1987.)
4. What are the advantages and disadvantages of panels compared to single surveys? (See Moser and Kalton, 1971, pp. 137–43.)

FURTHER REFERENCES

Bogart, L. (1967). No opinion, don't know, and maybe no answer, *Public Opinion Quarterly*, **31**, 331–45.

Hakim, C. (1982). *Secondary Analysis in Social Research*, Allen and Unwin, London.

McDaniel, S. W., Madden, C. S., and Verille, P. (1987). Do topic differences affect survey non-response results? *Journal of the Market Research Society*, **29**, 55–66.

Miller, W. L. (1987). The British voter and the telephone at the 1983 election, *Journal of the Market Research Society*, **29**, 67–82.

Assignments in Applied Statistics
Edited by S. Conrad
© 1989 John Wiley & Sons Ltd

Opinion Polling and Opinion Poll Designs

Stephanie Stray

School of Industrial and Business Studies, University of Warwick

OUTLINE

In the run-up to general elections the media report opinion poll figures indicating the level of support for the various political parties. However, many polling organizations conduct opinion polls on a regular basis, regardless of whether a general election is imminent or not, and seek information on a wide variety of topics and issues. This assignment describes two of the sample designs used by one polling organization and also gives an example of a typical questionnaire. The focus of the assignment is upon the type of sample designs used and their relative merits, and upon using and interpreting opinion poll data.

There are a variety of organizations in Britain that conduct opinion polls. Some of these undertake such polling regularly whilst others conduct surveys on a more *ad hoc* basis at times of specific political events, most notably during the run-up to a general election. The focus of this assignment is upon one particular polling organization (NOP) that conducts both regular monthly polls as well as additional polls at times when there is particular interest in British politics. The purpose of this assignment is to consider, in detail, two of the sample designs used by NOP as well as to examine an example of a typical questionnaire administered by them. Attention also focuses on the advantages and disadvantages of the sample designs.

KEYWORDS

opinion polls, sampling design, Random Omnibus Survey, interviewer bias, sampling interval, non-response, Quota Survey, interlocking quota, questionnaire design

Opinion polls are usually thought of in the context of elections and the prediction of election outcomes. However, this accounts for only a tiny fraction of the information and data that are collected by pollsters. They are also concerned with other aspects of the political climate including the assessments by the public of the political parties, the government, the Prime Minister, the leaders of the opposition parties and the influence of the government upon various aspects of the economy such as inflation and employment. Often polls are conducted at the time of specific political events such as party conferences and following the Chancellor's budget announcements in order to provide information for the media. However, political aspects are not the only subject matter considered by the pollsters. Opinion polls also seek to collect information upon a wide variety of social matters ranging from such topics as football and football hooliganism to tax evasion, defence and the armed services to religion, and kidney donorship to test-tube babies. Apart from seeking the views and opinions of the general public they also seek to determine the attitudes of specific subgroups of the population, such as trade unionists, magistrates, the police and civil servants. For a description of the difference between the functions of opinion polls and the functions of surveys in general together with details of some of the problems associated with interpreting opinion polls, see Marsh (1982, pp. 125–46).

The oldest British polling organization is Gallup which was established by Durant and which produced its first set of results in 1937. Gallup publishes its results in *The Daily Telegraph* and also in a regular monthly publication, *The Gallup Political Index*. Some years after Gallup's commencement of polling in Britain they were followed by National Opinion Polls (NOP) which began as a subsidiary of Associated Newspapers. Polling by NOP began early in 1957 on an *ad hoc* basis for *The Daily Mail* but did not start on a regular basis until 1961 from which time polls have been conducted each month. *The Daily Mail* no longer publishes NOP poll results on a regular basis although NOP ask the same basic set of questions weekly in their Random Omnibus Survey and the results are published in the *NOP Review*. Any client is permitted to buy space on the Random Omnibus Survey at a fixed rate per question. NOP remain one of the few polling organizations that use random, rather than quota, sampling on a regular basis.

After the establishment of these two polling organizations in Britain, a number of other companies also began to conduct opinion polls. These include Marplan, Opinion Research Centre, Research Services Ltd, Market and Opinion Research International (MORI) and Harris.

What follows is a description of two types of poll conducted by NOP, the first being their Random Omnibus Survey and the second their Quota Survey. The latter is used particularly at those times when a set of results is needed quickly, such as in the case of a by-election or in the run-up to a general election. Problems can occur with this type of sampling design, apart from the lack of standard errors associated with point estimates of population parameters, since despite the specification of the types of individuals that interviewers should select (interviewer quotas) there is evidence of interviewer bias in the selection of potential respondents.

NOP RANDOM OMNIBUS SURVEY

Samples are selected for the Random Omnibus Survey each week with the exception of Christmas and Easter. The procedure that is used is to first divide Britain's Parliamentary constituencies into nine groups corresponding to the Registrar General's Standard Regions for England and Wales (Northern, North-West, Yorkshire and Humberside, East Midlands, West Midlands, East Anglia, Wales, South-East, South-West) plus Scotland. Scotland is subdivided into the new Strathclyde Region and the rest of Scotland. The South-East Region in England is also subdivided into GLC (Greater London Council) and non-GLC constituencies and Wales is subdivided into South-East Wales and the rest. The next step is to subdivide the constituencies in each regional grouping into one of four types as follows:

1. *Constituencies in metropolitan counties.* These are constituencies that are contained completely within the metropolitan counties of Britain. In a few instances constituencies lie only partly within a metropolitan county and in these cases the constituencies are classified as belonging to this group only if 50 per cent or more of the population lies within the metropolitan area. In cases where a Standard Region contains two metropolitan counties (e.g. the North-West) the constituencies in each of these metropolitan counties are treated as two separate subgroups. The GLC constituencies are also subdivided according to whether they are north or south of the river and whether they lie to the east or west.
2. *Constituencies in urban administrative areas.* These groups are of constituencies which are composed only of urban local authority areas.
3. *Constituencies in which most of the population live in urban administrative areas.* These groups comprise constituencies consisting of both urban and local authority areas where the population density is 1.5 electors per hectare or greater.
4. *Constituencies in which most of the population live in rural administrative areas.* These groups consist of constituencies in both urban and rural local authority areas where the population density is less than 1.5 electors per hectare.

The above subdivision of constituencies results in 46 categories, as shown in Table 1, since not all four types of constituencies exist in each

Table 1
Groupings for NOP's Random Omnibus Survey (*Reproduced by permission of NOP Market Research Ltd*)

STANDARD REGION	SUBCATEGORY	NUMBER
Scotland	New Strathclyde Region	
	Glasgow	1
	Rest of Strathclyde	
	Urban	2
	Mixed	3
	Rural	4
	Rest of Scotland	
	Urban	5
	Mixed	6
	Rural	7
Northern	Tyne and Wear MC	8
	Urban	9
	Mixed	10
	Rural	11
North-West	Greater Manchester MC	12
	Merseyside MC	13
	Urban	14
	Mixed	15
	Rural	16
Yorkshire and	South Yorkshire MC	17
Humberside	West Yorkshire MC	18
	Urban	19
	Mixed	20
	Rural	21
East Midlands	Urban	22
	Mixed	23
	Rural	24
West Midlands	West Midlands MC	25
	Urban	26
	Mixed	27
	Rural	28
East Anglia	Urban	29
	Mixed	30
	Rural	31
Wales	South East	
	Urban	32
	Mixed	33
	Rural	34
	Rest	
	Mixed	35
	Rural	36
South-East	GLC North of River	
	East	37
	West	38
	GLC South of River	
	East	39
	West	40
	Non-GLC	
	Urban	41
	Mixed	42
	Rural	43
South-West	Urban	44
	Mixed	45
	Rural	46

region. Within each of these 46 groups the constituencies are listed according to the ratio of the Conservative to Labour vote at the previous general election. The electorate of each constituency is then listed and a cumulative total of the electorates is produced down the list of constituencies in each group. The next step is to select 180 constituencies from the groups in the following manner. The total number of (cumulative) electors on the list is divided by 180 to give a value f. A random number between 1 and f is selected. This random number is used to identify an elector in the list of cumulated electors. The *constituency* to which this elector belongs is thereby considered to be selected into the sample. To obtain the remaining constituencies the sampling interval f (as defined above) is added to the random number that was chosen to identify the first constituency in the sample. This gives a second number which identifies a second elector in the cumulated list of electors and hence identifies the second constituency. The procedure continues, by repeated additions of the sampling interval f to the number identifying the previously selected constituency, until 180 constituencies have been selected. Each constituency therefore has a chance of being selected that is proportional to the size of its electorate.

The consequences of this method of selection are that the sample contains a representative cross-section of urban and rural seats as well as having a mixture of safe Conservative, safe Labour and marginal seats. Note that only British constituencies are used and that Northern Irish constituencies are excluded. The 180 constituencies that are selected by this method are not reselected each week but are used for a period of time before a new set of 180 are chosen.

The next step is to select individual respondents for interview using the Electoral Registers of the selected constituencies. For each of the 180 constituencies a random number is chosen between 1 and E, where E is the number of electors in the constituency being considered. The random number identifies a particular elector on a numbered list for that constituency and also the polling district to which that elector belongs. This selected elector forms the starting point for selecting a group of electors from the Electoral Register. To select the group of electors every fifteenth elector on the Register is chosen from this starting point until eighteen names (twenty in Greater London) and their associated addresses have been selected. These names and addresses are supplied to interviewers in the appropriate geographical location.

The result of the above procedure is to make the sample representative of people on the Electoral Register. However, some people will have moved since the Register was compiled and others are not on the Register at all. The information for the Electoral Registers is collected in October of each year and is published the following February. By that time some 4 per cent of names are ineffective due to deaths and movers and this rises to about 16 per cent before the new Register is issued the following February.

In order to obtain a sample that is representative of all adults the interviewer is given an additional task over and above that of interviewing (if possible) the specified people that have been allocated. This task is to determine, for each address that is visited, whether the household contains anyone aged 15 years or over who is not on the Electoral Register. In urban areas the electoral register is in address order and interviewers are given a copy of the appropriate part of the Electoral Register so that they can check if a particular individual in the household is on the Register or not. In rural areas, where the Electoral Register is in name order, the interviewer has to rely on a member of the household to identify people not on the register. In either case, if one such person is found at an address they are interviewed, if possible, in addition to the person who was selected from the Electoral Register. If more than one such person is found then the interviewer lists the names of all non-electors alphabetically by surname and then by first name within surname. A random selection of the non-electors is made using a predetermined procedure in order to select a single non-elector to interview in addition to the person selected at that address from the Register.

Table 2 gives an example of how the selection of a non-elector is made. For example, at the fourth household to be interviewed, in a particular constituency, the interviewer determines the number of non-electors in the alphabetical list (that has been constructed for that household) and consults a table similar to Table 2 to determine which person on the list should be interviewed. From Table 2 it can be seen that this means that for the fourth household the person to be interviewed would be the first person listed if there were two non-electors in the household, the second person listed if there were three non-electors in the household, the first person listed if there were four non-electors in the household and so forth. Interviewers are required to make a fixed number of calls in order to attempt to interview the selected person(s) at each address and are not allowed to accept substitute members of the household.

This procedure means that every elector who is listed on the Register has the same chance of being selected for inclusion in the sample. This is not true, however, for those individuals who are *not* listed on the Register. For these people their chance of being selected into the sample depends upon two factors. These factors are the number of electors listed on the Register at a non-elector's address and the number of people living at that address who are *not* listed on the Register. This is further explained below.

Given that every elector has the same chance of being selected into the sample, the chance of a non-elector being selected depends on whether or not the non-elector's address is selected. Addresses that have more electors

Table 2
Selecting non-electors for interview. (*Reproduced by permission of NOP Market Research Ltd*)

SAMPLED HOUSEHOLD NUMBER	NUMBER OF NON-ELECTORS IN HOUSEHOLD					
	1	2	3	4	5	6 OR MORE
1	1	2	2	2	4	4
2	1	1	3	3	5	5
3	1	2	1	4	1	6
4	1	1	2	1	2	1
5	1	2	3	2	3	2
6	1	1	1	3	4	3
7	1	2	2	4	5	4
8	1	1	3	1	1	5
9	1	2	1	2	2	6
10	1	1	2	3	3	1
11	1	2	3	4	4	2
12	1	1	1	1	5	3
13	1	2	2	2	1	4
14	1	1	3	3	2	5
15	1	2	1	4	3	6
16	·	·	·	·	·	·
17	·	·	·	·	·	·
18	·	·	·	·	·	·

listed on the Electoral Register have a greater chance of being selected than addresses with fewer electors—this follows from the fact that all electors have an equal chance of selection. If a non-elector's address is selected then the chance of the non-elector being selected for interview depends on the total number of non-electors in the household. If there is only one non-elector at a selected address then that non-elector is certain to be selected for inclusion in the sample. However, if there are, say, four non-electors at a selected address then they each have a one in four chance of being selected. Therefore, a non-elector has a chance of being selected that is proportional to the number of electors at his or her address and that is inversely proportional to the number of non-electors at the address. Consequently, the final sample needs to be weighted to correct for this differential probability of selection of non-electors. The weighting factor that is applied to each non-elector in the sample is the number of non-electors at their address divided by the number of electors at their address.

For a variety of reasons it is impossible to interview all people selected into the sample. These reasons include refusals as well as people who prove impossible to contact or who are away at the time of the fieldwork. NOP reports that younger men tend to prove particularly difficult to interview because they tend to be out at work or out for social reasons. Hence the resulting samples that are achieved typically tend to over-represent older people and women. For further details on the nature of non-response in random opinion poll surveys, see Teer and Spence (1973, pp. 31–4, 57–60). To correct for this non-response problem the results are weighted each week, the weightings being based on a matrix that corrects the sample for the proportion of respondents in each age, sex and class group within each region.

Q1   Draw a diagram or chart of your own devising to illustrate the process used by NOP to select a sample for their Random Omnibus Survey. Mark on the diagram any salient features of the design.

Q2   What type of design do NOP use for their Random Omnibus Survey? What stratification factors, if any, do they use? How many stages are there in their sample design and what are these stages?

Q3   What advantages and disadvantages does NOP's Random Omnibus Survey design have over a simple random sample, of the same size, of the electorate for Britain as a whole?

Q4   Show that for NOP's Random Omnibus Survey each person on the Electoral Register has the same chance of being selected for inclusion in the sample.

Q5   How does the chance of selection of people on the Electoral Register differ from the chance of selection of people not on the Electoral Register for NOP's Random Omnibus Survey? What problems might this cause for the overall representativeness of the sample?

NOP QUOTA SURVEYS

In the case of quota sampling, interviewers are told to interview a certain number of people having specific background characteristics. They are not given the names and addresses of a specific set of individuals to contact but are free to choose whoever they wish, providing the people fit the background characteristics given to the interviewer.

NOP conduct quota samples for *The Daily Mail* and for *The Observer* based on people in 54 Parliamentary constituencies. The 54 constituencies are selected to ensure that the number of constituencies in each Standard Region is in proportion to the electorate in each of the Standard Regions. Furthermore, the constituencies are also selected so that the preceding general election's voting results of the selected constituencies, when

Table 3
An example of an interviewer's quota for NOP's Quota Surveys.
(*Reproduced by permission of NOP Market Research Ltd*)

PLEASE INTERVIEW 20 PEOPLE ACCORDING TO THE
FOLLOWING CONTROLS:

	MEN			WOMEN		
CLASS	ABC1	C2	DE	ABC1	C2	DE
AGE: 18–34	1	1	1	1	2	1
AGE: 35–54	1	1	1	1	1	2
AGE: 55 +	0	2	1	0	1	2

QUOTA NUMBER: __10__
CONSTITUENCY: _____
SAMPLING POINT NUMBER: ____

YOUR TEAM LEADER IS: _____
TELEPHONE NUMBER: _____

PLEASE RETURN THIS QUOTA SHEET TO YOUR
SUPERVISOR AND MARK ON IT THE NUMBERS
ACHIEVED IN EACH SECTION OF THE QUOTA

aggregated, are representative of the preceding general election results of Britain as a whole.

Interlocking quotas are used to determine the number of people to be interviewed. These quotas are based on age, sex and class such that, typically, there are eighteen cells that result (three age groups by three social class groups for both males and females). Interviewers are supplied with individual quotas such that when the quotas for all interviewers are aggregated then the resulting sample is representative of the population in terms of the background characteristics used for the quotas. An example of a quota given to an individual interviewer is given in Table 3.

Once interviewers have completed their quotas of interviews then the data are sent back to NOP by post, train or telephone depending on the speed with which the results are required. Weighting of results may well be required depending on the success individual interviewers have had in completing their quotas and depending on whether or not specific groups of the population have been over- or under-sampled. For further details on the representativeness of quota samples and the influence of interviewers on the established quotas, see Glenn (1970). There is obviously a delay between the fieldwork dates and the date on which a poll result is published and any interpretation of the results should bear this fact in mind.

Q6 What are the main features of the Quota Surveys used by NOP? What attempts are made to ensure a representative sample?

194

Table 4
Typical NOP questionnaire. (Reproduced by permission of NOP Market Research Ltd)

NOP/5782
(1–5)

SERIAL NO: _____ (6–9)

CONSTITUENCY NO: _____ (10) (11)

NAME: _____

ADDRESS: _____

TELEPHONE NO: _____

OCCUPATION OF HEAD OF HOUSEHOLD: _____

WORKING STATUS	(12)
Men working full-time	1
Men not working full-time	2
Women working full or part-time	3
Women not working	4

TRADE UNION MEMBER (13)

Yes ------------------ 1
No ------------------- 2

	ABC1	C2	DE	
18–34	1	2	3	(14) (12)
35–54	1	2	3	(15) /
55 +	1	2	3	(16) 16

Q1a How would you vote if there was
a general election tomorrow?
IF UNDECIDED OR REFUSED ASK Q1b

Q1b Which party are you most inclined to support?

	Q1a (17)		Q1b (18)	
Conservative	1	GO	1	ASK
Labour	2	TO Q2	2	Q2
Liberal/SDP	3	GO TO	3	GO TO
Other	4	Q3a	4	Q3a
Would not vote	5	GO TO Q4	5	GO
Undecided	6	ASK	6	TO
Refused	7	Q1b	7	Q4

Q3b Which party do you think you might switch to?
(21)

Conservative ----- 1
Labour ----- 2
Alliance ----- 3
Other ----- 4
Don't know ----- 5

ASK ALL
Q4 Regardless of how you may vote, who do you
think will win the next general election?
(22)

Conservative ----- 1
Labour ----- 2

ASK CONSERVATIVE AND LABOUR VOTERS ONLY

Q2 Would you consider voting for the Alliance if you felt that would prevent a party that you didn't like winning?

(19)

Yes - - - - - - - - 1 NOW
No - - - - - - - - - 2 ASK
Don't know - - - - 3 Q3a

ASK ALL NAMING A PARTY AT Q1a OR Q1b

Q3a How likely or unlikely are you to change your mind about who to vote for between now and the next election?

(20)

Very likely to - - - - - 1 ASK
Fairly likely to - - - - 2 Q3b
Fairly unlikely to - - - 3 GO
Very unlikely to - - - - 4 TO
Certain not to - - - - 5 Q4
Don't know - - - - - 6

Alliance - - - - - 3
No overall majority - - 4
Don't know - - - - - 5

Q5 When do you think Mrs Thatcher should step down as leader of the Conservative Party?

READ OUT (23)

Before the next general election - - - 1
After the next general election - - - 2
Not until the general election after that - - - - - - 3
Don't know - - - - - - 4

Q6 If the government continues with its present policies do you think unemployment will start to go down or not?

(24)

Yes - - - - - - 1
No - - - - - - 2
Don't know - - - - 3

Table 4 (*Continued*)

Q7 By the time of the next General election in 1988 do you think you personally will be better or worse off than you are now?

(25)

Better	1
Worse	2
Same	3
Don't know	4

Q8 *SHOW CARD A* There has been much discussion recently about possible changes in the rates. Which of these do you favour?

(26)

Local income tax	1
Flat rate tax	2
Property tax	3
Flate rate and property	4
Leave it as it is	5
Don't know	6

Q9 Who do you think has more influence in the Labour Party—Neil Kinnock or Arthur Scargill?

(27)

Kinnock	1
Scargill	2
Same	3
Don't know	4

Q12 I would finally like to ask you some questions on a different topic. Now that the £1 coin has been in use for some time, do you think the government was right or wrong to phase out the £1 note?

Right	1
Wrong	2
Don't know	3

Q13 *SHOW CARD C* The Royal Mint is considering these changes to the coinage. Overall do you think they are a good or bad idea?

(33)

Good	1
Bad	2
Don't know	3

Q14 Do you think these changes would make prices go up or down, or would they make no difference?

(34)

Up	1
Down	2
No difference	3
Don't know	4

Q10a Do you think the Labour Party is united or divided at the moment? *CODE BELOW*

Q10b Do you think the Conservative Party is united or divided at the moment? *CODE BELOW*

Q10c And do you think the Liberals and the Social Democrats are united or divided at the moment?

	Lab (28)	Con (29)	Lib/SDP (30)
United	1	1	1
Divided	2	2	2
Don't know	3	3	3

Q11 *SHOW CARD B* Are there any policies on this card which would make you definitely *not vote for* any party that was committed to them? Which?

(31)

No—none	1
Yes—ban clear	2
—abolish SERPS	3
—incomes policy	4
—repeal union laws	5
Don't know	6

Q15 If there was a new 10p piece the size of the old sixpence, do you think you might confuse it with the 20p piece or not?

(35)

Might	1
Not	2
Don't know	3

INTERVIEWER NAME: _____

INTERVIEWER NUMBER: _____

Q7   What are the relative advantages and disadvantages of the Quota Surveys compared to the Random Omnibus Survey used by NOP? (See Moser and Kalton, 1971, pp. 127–36.)

OPINION POLL QUESTIONNAIRES

Questionnaire design continues to be perhaps the least scientific part of the survey or polling process. Question wording, question sequence, interviewer effects and coding difficulties can all lead to biases in the results that are obtained.

An illustration of one of these effects is shown by some research conducted prior to the 1975 Referendum—about whether or not the United Kingdom should remain a member of the EEC—by NOP. Amongst several of the question formulations they used, two were 'Do you accept the government's recommendation that the United Kingdom should come out of the Common Market?' and 'Do you accept the government's recommendation that the United Kingdom should stay in the Common Market?' The difference between the percentages of people who were in favour of and those who were against staying in the EEC was 0.2 per cent for the former question and 18.2 per cent for the latter. (See Marsh, 1982, pp. 131–2, 141–5, for more details and for further examples.)

An example of a typical questionnaire used by NOP is given in Table 4.

Q8   Explain in detail how you would use the data collected from the questionnaire to estimate the percentage of the population indicating that they intend to vote for a particular political party? What does your answer suggest, if anything, about such figures that are produced by different polling organizations? (See Webb, 1980.)

Q9   What sort of biases in survey results can be caused by the questionnaire design? Which of these are relevant to the above questionnaire, and why? What sort of biases can be introduced by interviewers and what effect might this have on the estimation of the proportion of people voting for a particular political party? (See Cantril, 1944, pp. 108–10; Oppenheim, 1966, pp. 24–48; Moser and Kalton, 1971, pp. 270–82, 303–9, 423–8; King and Warren, 1980; Sudman, 1980.)

Q10   Of what use to you would the questionnaire be if you were a member of the Royal Mint and were considering changes to the coinage?

Q11   Comment on the overall satisfactoriness of the questionnaire and state any problems you feel exist. What changes would you make, if any, to overcome these problems?

Q12   Thursday, 11 June 1987 was polling day for a United Kingdom General Election. Consult newspapers within the three weeks prior to that date to determine the results for any two opinion polls published *on the same day* that indicated the way the electorate intended to vote at the election. Record any details that are given of the sample designs that were used and the methods by which the polling organizations collected the data. What differences exist in the results for the two polls you have chosen? Why would you expect the results to differ and what factors may have given rise to these differences?

REFERENCES

Cantril, H. (1944). *Guaging Public Opinion,* Princeton University Press, Princeton, New Jersey.

Glenn, N. D. (1970). Problems of comparability in trend studies with opinion poll data, *Public Opinion Quarterly,* **34**, 82–91.

King, A. L., and Warren, M. (1980). Survey on opinion, attitude and voting during the 1979 British election campaign. In *Opinion Polls,* ESOMAR, Netherlands, pp. 231–50.

Marsh, C. (1982). *The Survey Method,* Allen and Unwin, London.

Moser, C. A., and Kalton, G. (1971). *Survey Methods in Social Investigation,* Heinemann, London.

Oppenheim, A. N. (1966). *Questionnaire Design and Attitude Measurement,* Heinemann, London.

Sudman, S. (1980). Reducing response error in surveys, *The Statistician,* **29**, 237–73.

Teer, F., and Spence, J. D. (1973). *Political Opinion Polls,* Hutchinson, London.

Webb, N. L. (1980) A note on the accuracy of forecasting opinion polls, *Journal of the Royal Statistical Society (Series A),* **143**, 367–8.

SUPPLEMENTARY QUESTIONS

1. Consider the sample questionnaire used by NOP. If you were advising the leader of one of the political parties in Britain about the support for their party at a forthcoming election and about the nature of their support, what analysis would you wish to have performed upon the data collected by the questionnaire?

2. What are the major problems associated with interpreting opinion poll data? (See Marsh, 1982, pp. 125–46.)
3. What are the main advantages and disadvantages of employing stratification in a random sample? (See Moser and Kalton, 1971, pp. 85–100.)
4. What are the main advantages and disadvantages of employing a multistage design in a random sample? (See Moser and Kalton, 1971, pp. 100–16.)

Assignments in Applied Statistics
Edited by S. Conrad
© 1989 John Wiley & Sons Ltd

Technology and the Survey Process

Stephanie Stray

School of Industrial and Business Studies, University of Warwick

OUTLINE

Current developments in technology are leading to changes in the nature of the survey process. The advent of comparatively cheap microcomputers has enabled modifications to be made to data collection methods and to data checking. This assignment begins by considering the stages that need to be undertaken in the traditional survey process and then turns to some of the developments and changes that are being introduced as a result of the influx of technology. Attention focuses upon three different types of data collection procedures and the alterations to the survey process that have resulted. These data collection procedures are computer assisted personal interviewing (CAPI), computer assisted telephone interviewing (CATI) and computer assisted panel research (CAPAR). A recent example of the Dutch experience of this latter form of survey research is described.

KEYWORDS

computer technology, data collection, pilot study, fieldwork, coding, microcomputers, computer assisted personal interviewing (CAPI), interviewing schedule, questionnaire routing, interview error, computer assisted telephone interviewing (CATI), weighting scheme, computer assisted panel research (CAPAR), panel data, diaries, questionnaire design, response rates

The current trends in the development of widespread computer technology are now leading to changes in the survey process and in particular to changes in, and departures from, the traditional data collection methods of mailed questionnaires and face-to-face interviewing. These changes have arisen not only because of technological developments but also because of the increasing cheapness of computer hardware and the availability of portable microcomputers and microcomputers for home use. The implications and ramifications of these changes are less well documented than those of the more conventional data collection methods for obvious reasons and, consequently, a number of issues arise that are centred around the need to explore and to evaluate the effects of these changes.

This assignment is concerned with some of the new developments in survey methodology, and in data collection methods in particular, that have arisen as a result of technological change (see Jackling, 1984). First, a description is given of the traditional survey method process and this is then followed by details of three new data collection procedures that are currently in use. Finally, an example is given of the recent implementation of technology in panel research by the Gallup organization in Holland.

THE TRADITIONAL SURVEY METHOD PROCESS

The traditional survey method process involves a number of stages and requires statistical expertise, not only at the analysis stage but also in the sample design, which must additionally be coupled with organizational skills. Funds are rarely, if ever, unlimited and consideration needs to be given to the accuracy of results, the cost, time and labour involved and how best to deploy resources. The initial stage involves both the study of the existing literature and statistical data as well as discussions with experts in the field to be studied. There is often little point in duplicating work that has already been done or in collecting new data when information already exists to answer the questions at hand. However, a clear specification must be determined of the types of people to be studied and the data that are required. For instance, it is of no use to say that one wishes to determine purchasing patterns for clothes by young people without specifying exactly what is meant by a 'young person' and what sort of clothes purchases one is considering—clothes for work, clothes for sporting activities, clothes for leisure purposes, etc.—as well as a more precise definition of 'clothing' (does it include shoes, for example?).

Assuming resources are available to enable only a limited coverage of the population (i.e. there is a need to select a sample) then questions arise as to what type of sample is to be used, who is to be sampled (individuals, households, shops, etc.), what sampling frame is available, how large a sample is required, how to minimize non-response (the use of follow-up interviews and call-backs for instance), what form of data collection to use (personal interview, mailed questionnaire and so forth), the timing of the data collection and the resources that are required to process the survey data. When these issues have been addressed it is very likely that a questionnaire of some form will be required and hence consideration needs to be given to its design, question wording, question sequence, layout, the associated instructions and printing, and the details of the analysis to be undertaken on the resulting data. Collecting data with no clear idea of how it is to be analysed is likely to be a waste of both time and money.

One also needs to plan the organization of the fieldwork. If interviewers are to be used, how many are required and where are they to be located? If mailed questionnaires are employed who is going to distribute them? Neither can one exclude consideration of what is to happen to the questionnaires once they have been completed. In most instances the responses provided by the individuals who have cooperated in the survey will need coding for data entry for computer analysis as well as checking for omissions and inconsistencies. Manpower is obviously required for these tasks as well as computer power.

Before launching into the survey itself it is vital that a pre-test or pilot study be undertaken. This enables one to test out the questionnaire on a limited number of people (who should be similar to the potential respondents) in order to check that the questions are understood and that respondents can provide the necessary information. However, the pilot study should also be designed to test all the features of the main enquiry, and this includes the organizational design and running of the survey as well as the more statistical aspects.

Only after all the above steps have taken place should the next stages commence, which include the final printing of the documents (question-naires, questionnaire instructions, interviewer instructions, etc.) and the actual selection of the sample of respondents. Once the respondents have been selected they can then be contacted by post or by interviewer in order to secure their cooperation and participation in the survey. Having collected the information from the respondents it then needs to be sent back to central headquarters for coding, checking and processing, and for the analysis to be undertaken. Finally the results of the analysis need to be presented.

Q1 Draw up a chart or diagram of your own devising to represent the various stages of the traditional survey process that use face-to-face interviews. The diagram should show the various stages in the sequence in which they occur. (See Moser and Kalton, 1971, pp. 47–51; Oppenheim, 1966, pp. 1–23.)

The following descriptions concern three different approaches that are currently in use that result in modifications to the above process. All of these approaches are concerned with changes in the method of data collection although this is not to say that they do not have implications for other aspects of the survey process.

COMPUTER ASSISTED PERSONAL INTERVIEWING (CAPI)

The typical CAPI system involves providing interviewers with portable microcomputers. The interviewer contacts a central computer at the headquarters of the survey institution via telephone and, using a modem, downloads a copy of the questionnaire which is stored directly on the interviewer's micro. Also supplied is a list of the names and addresses of the people to be contacted for interview. The interviewer contacts these people in the normal manner in order to conduct the interviews. Instead of reading the questions from the questionnaire and recording responses to the various questions on the interviewing schedule the interviewer uses the portable micro to enable him or her to conduct the interview. The micro contains software which, when run, produces each question to be asked, in turn, upon the screen. The response given by the interviewee—who either has the questions on the screen read out or reads the questions directly—is typed by the interviewer directly into the micro. Appropriate software on the micro causes these responses to be stored. All questionnaire routing (moving to the next appropriate question to be answered) is automatically undertaken by the micro.

For instance, suppose part of the questionnaire is concerned with travel to and from one's place of work—the means of travel, time taken and so forth. One first needs to ascertain if the respondent has a job since if not then there is obviously no point in asking the subsequent questions about travel to and from work. Therefore one of the questions put to the respondent, prior to asking about travel to and from work, will be designed to determine if the respondent is currently in employment. If the code that is entered in to the micro indicates that the response is negative, the micro will be programmed to automatically skip all the remaining questions concerning travel to and from work. A conventional interview would entail having instructions on the recording schedule that indicated to the interviewer that if the respondent did not have a job then the interview should continue by omitting the questions relating to travel to and from work and proceeding directly to the appropriate next question on the next topic. Obviously such questionnaire routing can become quite complex and, if left to the interviewer, can become a potential source of interview error.

The micro can also be programmed to check for any inconsistencies that may arise in the responses given by the interviewee or due to misrecording by the interviewer. In the former case it is obviously far more efficient for the micro to do this than for the interviewer to attempt to cross-check responses, especially in a long interview. In the latter case any such errors may well not be detected until checks are performed on the data back at the central headquarters of the organization and by that time it may well be too late to correct such errors.

Once the interviews have been conducted the interviewer again makes contact with the central computer via telephone and the central computer system records all the data that the interviewer has collected and stored on the micro.

Q2   What differences to the traditional survey process occur as a result of the CAPI system? How might these differences affect the overall cost of a survey?

Q3   What are the relative advantages and disadvantages of the CAPI system over traditional face-to-face interviewing?

COMPUTER ASSISTED TELEPHONE INTERVIEWING (CATI)

Telephone interviewing is a relatively recent form of data collection in Britain although it has a much longer history in the United States. The major criticism of this form of data collection concerns the possible bias that may creep into the sample because of confining the sample selection to telephone owners.

Nevertheless, telephone interviewing continues to become more widespread within Britain and computerization can also aid in reducing some of the workload associated with this type of procedure. (See Miller, 1987, for a description of some of the biases in telephone interviewing as well as details of a weighting scheme designed to reduce the bias of this form of sampling.) A typical CATI system would be for a central computer system to down-load the names and telephone numbers of the people to be interviewed to a micro as well as details of the questionnaire to be used. The interviewer consults the micro to obtain details of who is to be interviewed and then contacts each of the respondents via telephone. Interviews are conducted immediately over the telephone or appointments are made for an interview (over the telephone) at a time that is convenient for the respondent.

The interviewing session is, as in the CAPI system, controlled by the micro which indicates the questions to be asked of the respondent. The interviewer reads the questions to the respondent over the telephone and types the interviewee's responses directly into the micro where they are stored. As previously, at some later point the resulting data are transferred back to the central computer system.

Q4 What changes result to the diagram you have drawn in your answer to Question 1 if conventional telephone interviewing is used? How might these changes affect the overall survey cost?

Q5 How does the CATI system further modify the changes to the survey process and the survey cost that you have indicated in your answer to Question 4? (See Nicholls and Groves, 1986.)

COMPUTER ASSISTED PANEL RESEARCH (CAPAR)

This particular procedure, as its name suggests, involves the use of a panel of individuals whose members are contacted for interview on successive occasions. Traditionally this involves repeated trips by an interviewer to collect data from the respondent at different points in time or the repeated completion of mailed questionnaires by the respondent. Also, diaries are extensively used in panel research whereby the respondent keeps a record of daily activities, purchases or whatever is required and periodically sends the completed diary (for a particular week or month) to the survey organization. Consequently, one can use these diaries to monitor household purchasing and expenditure patterns as well as examining the effectiveness of, say, regional advertising campaigns.

Unlike the two other procedures described above the CAPAR system does not involve supplying the interviewer with a micro but requires the panel member to be supplied with a micro directly. The panel member then uses the micro in his or her own home environment. Each week or each month—or however often is required—the panel member contacts the survey organization via telephone and a questionnaire is transferred directly on to the panel member's micro. The micro also contains appropriate software to enable the panel member to go through the questionnaire and to enter responses to whatever questions are asked. The responses are stored on the micro as in the previous approaches. The panel member is, typically, given a certain time by which they should have completed the questionnaire and then needs to contact the central organization by telephone in order that their data may be transferred on to the central system.

One particular advantage of this system is that any poorly phrased questions or mistakes in the questionnaire design can be picked up early in the data collection process and any necessary changes can be made to the questionnaire before issuing it to further members of the panel. Since not all panel members will contact the central organization at the same time in order to receive their questionnaires any problems encountered by the early respondents do have an opportunity for correction.

An additional advantage of such computer-based panel research is that information provided by a respondent during one 'interview' can easily be used in subsequent 'interviews' since it can be stored either centrally or on the home micro. This avoids the need to ask repeatedly for data upon background characteristics of the respondent as well as other data which may be required for the analysis of responses to questions asked in later interviews. It is much more tedious to make use of such previously collected data if traditional pen and paper methods are used for recording respondents' answers. Furthermore, for time/budget research where the respondent is required to complete a diary of activities during the course of the day it is arguable that more accurate or more detailed information can be gained. In the traditional diary system it is perfectly possible for an individual to say that they spent from 8.15 to 9.00 in the morning having breakfast and then record that at 9.00 a.m. they were doing household shopping. In a computer-based system whereby the individual 'accounts' for the day by responding to a computer-based interview such entries can be queried at the time of completion since some time must have been spent travelling (from a home-based activity) to the shops in order to do the shopping (which is a non-home-based activity). One can therefore ascertain at what time the individual left home to go to the shops, how long the journey took and, if required, the means of travel.

An example of a CAPAR system

The Sociometric Research Foundation in Holland has developed an example of a CAPAR system and this is now used by the Dutch Gallup Institute, NIPO. In March 1986 a random sample of 500 households in Holland was selected and in June of the same year this was extended to a random sample of 1000 households. All members of the households aged 12 years and over formed a panel of 2300 individuals. Each household received a free home computer (Philips MSX2) and a modem (to transmit information between NIPO's central computer and the home micro via telephone). Also included was the necessary software to deal with communications between the micro and NIPO's central computer system, to administer the questionnaires and record responses and to determine who in the household has or has not answered the questionnaires. Households selected for inclusion in the panel who did not own a television set were provided with once since the micro is linked to an ordinary domestic television to provide the screen to display the questions.

Answers to the questionnaires are given by choosing a numeric category, choosing a category on a rating scale, specifying a numerical answer or by drawing lines where the line length indicates a respondent's strength of preference for an item or the extent of their agreement with a statement (Saris, 1982; Doorn, Saris and Lodge, 1983). Open-ended questions where the respondent is free to type a reply are also possible.

At the beginning of each weekend a member of each household contacts NIPO by telephone and, using the modem, a questionnaire is downloaded onto the household's micro. The software on the micro gives instructions as to which member or members of the household are required to complete the questionnaire during the course of the weekend. Not all members of the household are necessarily asked to complete each questionnaire on every occasion—the actual number of respondents that is required varies according to the topics about which information is sought. Typically, each questionnaire takes about 30 minutes for a respondent to complete. The software on the home computer keeps track of the household members who have and who have not answered the questionnaire, and at the start of each 'interview' the household members who have yet to answer the questionnaire are listed on the screen by name as a reminder. At the beginning of the following week the data stored on the micro as a result of the responses entered by the relevant household members are sent, via modem and telephone, back to NIPO for analysis. Panel members who have not answered the questionnaire by the Monday following the weekend are telephoned and asked to complete the questionnaire if possible. Individuals deciding that they no longer wish to continue to participate in the panel are required to return their micro to NIPO.

NIPO also runs a telephone service for queries from panel members if they have problems with the hardware, software or a particular questionnaire. (See *The Observer*, 10 May 1987, for more details.) Telephone costs incurred by the households as a result of their participation in the panel are reimbursed by NIPO. Panel members are able to obtain computer games for use on their micro.

At the beginning of the study a large number of households were contacted and asked if they would be willing to cooperate in a new project. They were told of the longitudinal nature of the study but were not informed of the fact that they would be given a home computer. Of those agreeing to participate a random sample of 500 was selected such that the sample was representative of a large number of background characteristics of the Dutch population. Out of these 500 households that were contacted only 23 refused to participate once they had learnt about the actual procedure that was to be followed and these were replaced by households with similar background characteristics. For those households participating in the study, Table 1 shows the response rates for individuals for four

Table 1
Response rates for NIPO for four weekends in May and June 1986.
(*Reproduced by permission of W. E. Saris*)

	MAY		JUNE	
	3–4	24–25	14–15	28–29
Response rate on Monday	84.2	91.8	89.6	86.8
Response rate after recall	93.6	97.0	95.2	92.4
Non-response	6.4	3.0	4.8	7.6
Reasons for non-response				
Not at home	3.2	2.0	3.0	5.6
Technical problems	1.4	0.4	0.6	0.2
Moving	0.0	0.4	0.2	0.2

weekends in May and June 1986 as well as the effect of the Monday telephone call to remind people who have not replied to the questionnaire during the course of the weekend. After 12 months of using this system NIPO has encountered a cumulative panel dropout rate of 8 per cent compared to 40 per cent per year in their traditional panel studies. These figures appear to compare favourably both with the Social Economic Panel of the Central Bureau of Statistics in Holland which has a non-response rate of around 38 per cent and with the telephone panel of the Netherlands Broadcasting Organisation (NOS) which had an average non-response rate of 19 per cent in weekly interviews and a cumulative drop-out rate of 10 per cent over a year. Furthermore, NIPO claim that they appear to be getting more accurate data from respondents by using the CAPAR system rather than more traditional face-to-face interviewing techniques. In support of this they quote that in one questionnaire they asked panel members about their usage of drugs and pills and recorded a usage rate about 20 per cent higher than that found by previous studies.

This type of procedure obviously enables information to be collected and analysed very quickly. An illustration of this is provided by the analysis following a television debate prior to a recent Dutch general election. Panel respondents were asked to watch the debate and then to answer a questionnaire from NIPO. Within 30 minutes of the end of the televised debate the data from around 800 respondents had been collected and analysed and the results made available for a television news broadcast.

Q6 If respondents are selected to be part of a panel what stages of the survey process need to be repeated each time a questionnaire is issued to the panel?

Q7   What changes result to the panel research process as a result of the CAPAR system?

Table 2
Results of Sociometric Research Foundation Study. (*Reproduced by permission of W. E. Saris*)

	NUMBER	PERCENTAGE
Reasons for non-response		
in CAPAR survey		
Refused at second step		
when computer was introduced	14	9
Refused due to technical		
problems with computer	8	5
Refused to participate		
in panel because of		
Lack of time	21	13
No interest	23	15
Cooperated in panel	90	58
Total	156	100

The Sociometric Research Foundation has also undertaken its own experiments of the CAPAR system. One of these involved selecting a random sample of individuals and asking them to participate in a normal face-to-face interview. During the course of this interview respondents were asked whether or not they were willing to answer a second interview on a computer. Those who cooperated in this second stage and who went ahead with the second interview were asked to form a panel of respondents using the computer on a regular basis. Table 2 summarizes the outcome of this procedure. The characteristics of respondents were also noted so that comparisons could be made between those willing to participate in a face-to-face interview and those willing to cooperate in the panel. Details are given in Tables 3 and 4 of the numbers of respondents willing to cooperate by age group and by income group respectively.

Q8   What can be said about the response rates achieved by NIPO and by the Sociometric Research Foundation in their use of the CAPAR system? What problems or benefits, if any, does this suggest that this form of survey research might have?

Q9   What other aspects of the response rates do you feel ought to be examined and why?

Q10 What differences might result in questionnaire design from using the CAPAR system compared to the more traditional self-completion questionnaires?

Table 3
Age group of respondents cooperating in face-to-face interviews and in the panel. (*Reproduced by permission of W. E. Saris*)

AGE GROUP	FACE-TO-FACE INTERVIEW	PANEL
<35	40	25
35–54	64	44
55–70	41	18
>70	11	3
Total	156	90

Table 4
The differences with respect to income between the sample of the face-to-face study and the tele-interview panel. (*Reproduced by permission of W. E. Saris*)

INCOME	FACE-TO-FACE INTERVIEW		TELE-PANEL	
	NUMBER	%	NUMBER	%
<1500	21	14	11	13
1500–2500	66	44	40	46
2500–4250	48	32	27	31
>4250	14	10	9	10
Unknown	7		3	
Total	156	100	90	100

REFERENCES

Doorn, L. van, Saris, W. E., and Lodge, M. M. (1983). Discrete or continuous measurement: what difference does it make? *Kwantitatieve Methoden*, **10**, 104–21.

Jackling, P. (1984). Computer assisted questionnaire design: the real breakthrough. In *Are Interviewers Obsolete? Drastic Changes in Data Collection and Data Presentation*, ESOMAR, Amsterdam, pp. 23–5.

Miller, W. L. (1987). The British voter and the telephone at the 1983 election, *Journal of the Market Research Society*, **29**, 67–82.

Moser, C. A., and Kalton, G. (1971). *Survey Methods in Social Investigation*, Heinemann, London.

Nicholls, W. K., and Groves, R. M. (1986). The status of computer-assisted telephone interviewing: Part 1—Introduction and impact on cost and timeliness of survey data, *Journal of Official Statistics*, 2(2), 93–115.

The Observer (1987). Plug into the armchair poll (P. Spinks), 10 May, p. 39.

Oppenheim, A. N. (1966). *Questionnaire Design and Attitude Measurement*, Heinemann, London.

Saris, W. E. (1982). Different questions, different variables. In C. Fornell (ed.), *Second Generation Multivariate Analysis*, Praeger, New York, pp. 78–96.

SUPPLEMENTARY QUESTIONS

1. What are the reasons for undertaking a pilot study and what problems in the survey process may be detected by a pilot study? (See Moser and Kalton, 1971, pp. 47–51.)
2. What are the differences in the kinds of questions that can be asked in a face-to-face interview compared with a self-completion questionnaire? (See Moser and Kalton, 1971, pp. 256–349; Oppenheim, 1966, pp. 24–48.) What implications, if any, do these have for self-completion computer-based questionnaires?
3. The Social Survey Division of the Office of Population Censuses and Surveys conducts surveys upon a wide variety of topics. Consult your library to find out which of these are available to you. Choose one of the surveys that interests you and determine, for each of the stages that you have listed in your answer to Supplementary Question 1, the details of what had to be undertaken for the specific survey you are considering.

Assignments in Applied Statistics
Edited by S. Conrad
© 1989 John Wiley & Sons Ltd

The Analysis of Cross-Tabulated Survey Data

G. J. G. Upton

Department of Mathematics, University of Essex

OUTLINE

Problems in analysing survey data, where the data can be summarized in a multidimensional contingency table, are introduced. The analysis of such data employs log-linear or related models which enable the analyst to consider large numbers of variables simultaneously, as opposed to two at a time. The assignment concentrates on conveying ideas about associations between variables, including the fact that several associations may be present simultaneously and that interactions may involve three or more variables at once.

KEYWORDS

log-linear models, logit models, dirty data, contingency table, independence, collapsing a table, Simpson's paradox, ecological fallacy, interaction, hierarchy principle, factors, responses, path diagram, saturated model, structural zero, simultaneous testing

Previous assignments in this section have concentrated on various methods of collecting data. The principal themes have been the need to make a sample representative of the population of interest and the need to have the information available quickly for analysis by computer. However, there is no point in collecting data if we cannot analyse it and this assignment therefore concentrates on one possible approach which presupposes that the data can ultimately be summarized as a table (or set of tables) of frequency data or counts.

The last twenty years have seen an enormous increase in the power of computing equipment and this has brought with it a capability for data analysis of a type that would not previously have been feasible. A major growth area has been in the development of methods for analysing survey data using so-called *log-linear* and *logit* models. The intention of this assignment is to familiarize you with some of the ideas underlying these models (without actually becoming too involved in the underlying mathematics).

We shall refer to a set of data collected in an interesting experiment performed at Sussex University. The experiment is described in some detail later, but you may like the challenge of devising an alternative equivalent experiment.

Q1   Devise an experiment that will provide information to answer the following question: 'Is it true that, in this country, people are more helpful to females than to males?' (You should, in addition, suppose that age and race may also be relevant.)

Before revealing the Sussex answer to Question 1 we consider some more general questions. There are, in particular, four problems that can be expected with survey data:
1. The amount of data available for analysis always exceeds the capability of the available computer.
2. No data set ever contains all the information required.
3. In order to understand the data it will be necessary to spend many hours talking to the person who originated the survey.
4. The data will always be 'dirty'—that is there will always be 'impossible' values and contradictory combinations of responses.

Q2   List as many ways as you can that might lead to a set of survey data being described as 'dirty'.

A typical survey, such as the British Election Study survey, will collect information on a great many different issues. This is partly in order to avoid the second of the problems listed above (though it will still occur!). Also, of course, the marginal cost of collecting and recording information concerning the response to an extra question is comparatively small. Surveys typically err on the side of too many, rather than too few, questions and hence automatically produce problem 1.

In this assignment we do not have the space, and you do not have the time, to analyse the data from a major survey. Instead, therefore, we concentrate on the Sussex experiment that was designed in an attempt to answer Question 1.

The experiment is described in detail by Sissons (1981). In essence it involved eight similarly dressed students who approached people with a 5p coin in hand and said 'Excuse me, do you have change for 5p?'

Q3 Suppose that you were one of these students. What information would you record afterwards concerning the results of an interview?

Q4 Suppose you were arranging this experiment. How would you choose your students and what instructions would you give them?

TWO VARIABLES AT A TIME

The results from the Sussex experiment can be summarized as a set of counts that reflect the number of cases corresponding to various combinations of circumstances. We can summarize a set of survey data in exactly the same way and therefore the methods for surveys can be illustrated with this much simpler set of data. Prior to 1970, the standard procedure following such a summarization would involve a scrutiny of the data taken two variables at a time. The object of this procedure would be to discover which pairs of variables are related to one another. We shall commence our analysis in this old-fashioned way, and will begin by looking at the effect that the sex of the requestor has on the chance of being given help. The data from the Sussex study is summarized appropriately in Table 1, which is referred to as a *contingency table*.

If the giving of help is independent of the sex of the student asking for help, then the percentages being given help should not differ significantly.

Q5 Test the hypothesis that the table displays independence.

You will have deduced from the original question in this assignment that there is more to this data set than has so far been revealed. Here is a little more information! Of the eight students, four were Asian and four were

Table 1

STUDENT	HELP GIVEN	NO HELP GIVEN
Male	71	29
Female	89	16

Table 2

STUDENT	HELP GIVEN	NO HELP GIVEN
English	88	12
Asian	72	33

English. All eight students spoke English fluently. All the people who were approached and asked for change were English, all were in a shopping precinct and all were walking alone. Table 2 gives some more summary figures.

Q6 Have you any comments on this table?

Here are some more facts! There were two Asian male students, two English male students, two Asian female students and two English female students. Each student was supposed to interview 25 people, but one Asian female student interviewed 30 people. At first sight this would appear to unbalance the data, but in practice it makes very little difference.

Q7 Consider the 160 people who gave help. Using the totals given in the first columns of Tables 1 and 2 as marginal totals of a 2 × 2 table, construct a table that shows what counts we would have observed if the effect of a student's sex and the effect of that student's race had been independent of one another. In other words, how many English males would you have expected to have obtained help?

THREE VARIABLES AT A TIME

Your results so far will have suggested that both the race and the sex of the student asking for help had some bearing on whether that student was given help. However, these statements may be misleading since the variations in the helpfulness of the respondents may be due principally to the *combination* of the race and sex of the requestor. It is impossible to obtain information about the effects of such combinations using two-way tables of the kind considered so far. We therefore now consider the simultaneous information provided by Table 3.

Table 3

STUDENT	HELP GIVEN	NO HELP GIVEN
English male	41	9
English female	47	3
Asian male	30	20
Asian female	42	13

Q8   Compare the first column of Table 3 with the figures you calculated in Question 7. What do you conclude?

Q9   For each of the four categories of interviewer, calculate the percentages of interviewees that gave the students help. Comment on your results and, without performing any complicated statistical analysis, quantify the apparent effects of the race and sex of a student on his or her chance of obtaining help.

Q10   The data of the two-dimensional Table 1 could be represented as

$$71—29$$
$$| \quad |$$
$$89—16$$

which is a square with the cell frequencies arranged appropriately. Draw the corresponding figure that represents the data of Table 3.

In the present case we do not seem to have been misled by the results of 'collapsing' Table 3 over the sex of the student (to get Table 2) or over the race of the student (to get Table 1). However, it is certainly possible to get so seriously misled by collapsing a table that we get an entirely false impression. Consider, for example, the hypothetical data of Table 4.

Q11   Repeat Question 9 for the data of Table 4.

Q12   Collapse Table 4 over the sex of the student and, separately, over the race of the student. Calculate the percentages of interviewees obtaining help for the separate rows of these tables. Comment on the difference between your results here and those of Question 11.

Q13   Provide a graphical analogy for this phenomenon, using regression lines superimposed on a hypothetical scatter diagram showing the relation between two variables X and Y.

The phenomenon is known as 'Simpson's paradox' and its counterpart in regression is known as the 'ecological fallacy'. Most texts dealing with multidimensional contingency tables will provide examples of the problem, which can manifest itself in a wide variety of ways. Consider, for example, the hypothetical case where all English males and all Asian females receive

Table 4

STUDENT	HELP GIVEN	NO HELP GIVEN
English male	25	5
English female	75	15
Asian male	30	150
Asian female	10	50

help, whereas no English females and no Asian males receive help. If we collapse over race then we find that both sexes receive help about 50 per cent of the time and it therefore appears superficially that sex plays no part. The truth is, of course, that what really makes the difference is the *combination* of race and sex of the student.

If one variable affects another then we say that there is an association or *interaction* between these variables. If the interaction involves the joint effect of two variables on a third, then this is called a three-variable interaction. The modern approach to data involving several categorizing variables is to use models (termed log-linear or logit models) that specifically allow for the possible presence of all possible such interactions.

As might be expected from their names, these models work with logarithms of the observed counts, rather than with the counts themselves. The reason for the logarithms is that their presence enables the standard theory of regression and the analysis of variance to be extended to this rather different class of data. In the present assignment, however, the aim is not to replicate these models but to illustrate their ideas and terminology.

The log-linear model came into prominence in the early 1970s as a consequence of a series of papers by Goodman (see Bishop, Fienberg and Holland, 1975) that appeared in various sociology and statistics journals. The first definitive account of the subject is that of Bishop, Fienberg and Holland (1975). Those with only a reasonable grasp of basic statistics may prefer the introductory accounts by Upton (1978) or Fienberg (1980). A more modern, and remarkably comprehensive, account is provided by Fingleton (1984). For those who are nervous of anything at all mathematical, a words-only account is provided by Gilbert (1981).

For conciseness, in our discussion of the possible interactions, we denote the variables by:

A: Helpfulness
B: Race of student
C: Sex of student.

Thus far we have distinguished interactions between A and B, and between A and C, and we can denote these interactions as being the AB and AC interactions. Comparison of Tables 1 and 2 with Table 3 suggested that there was no significant evidence of the three-variable ABC interaction. This means that there was no evidence that the interaction that exists between helpfulness and the race of the student was affected by that student's sex.

We now want a succinct method of describing the properties displayed by a set of data. In other words, what (log-linear) model would provide an appropriate description of a set of data such as that of Table 3? Using the notation of Upton (1978) and Fingleton (1984), we can describe the model for this data set as being the model [AB/AC], where the symbol '/'

is used as a terminator rather than a divide sign. This model description is simply a list of the apparently relevant interactions included within square brackets. Question 14 tests your understanding of the notation.

Q14 The model [A] implies that helpfulness is unrelated to either the sex of the student asking the question or the race of that student. What does the model [AB] imply?

In addition to the interactions of interest our model definition should always include 'interactions' that have been fixed in advance, either by chance or, as in the case of the Sussex data, by design. For the data of Table 3 we should have included the BC interaction within our model definition because the total counts for the combinations of the categories of these variables were pre-ordained (at 50, 50, 50 and 55). We return to this point in more detail a little later, but at the present time we simply note that the appropriate model for Table 3 is not [AB/AC] but [AB/AC/BC].

Q15 Draw a tree diagram showing the five possible models, *all including* BC, that could be used to describe alternative data sets of the type illustrated in Table 3.

FOUR VARIABLES AT A TIME

It is now time to introduce one final level of complexity to the Sussex data by revealing that the students were instructed to approach roughly equal numbers of males and females. The resulting 2^4 table appears as Table 5.

Q16 Calculate the percentages giving help for the various student–respondent combinations. Comment on the results.

We can denote the fourth variable, sex of the respondent, by the letter D. There are therefore now numerous different kinds of interactions involving A with one, two or three of the other variables.

Table 5

| | RESPONDENTS | | | |
| | MALE | | FEMALE | |
STUDENT	HELP GIVEN	NO HELP GIVEN	HELP GIVEN	NO HELP GIVEN
English male	20	4	21	5
English female	23	0	24	3
Asian male	9	15	21	5
Asian female	25	2	17	11

Q17   How many possible interactions are there?

Also, of course, there is a huge increase in the number of possible models (from five to nineteen). We can have exotic models such as [AB/ACD/BCD], which states that the percentage giving help depends both on the race of the student and on a combination of the sex of the student and the sex of the respondent. However, note that the model [AB/ABD/BCD] would be written simply as [ABD/BCD], because if a model has taken account of the dependence of variable A on the combination of categories of variables B and D, then the dependence of variable A on variable B alone has already been accounted for. This is called the *hierarchy principle*.

Q18   Bearing in mind the hierarchy principle, draw a tree diagram to show the nineteen possible models, all including the BCD interaction, that are possible in this case. Which model appears to you to be most appropriate?

Suppose there is an interaction between two variables, X and Y, and suppose that it is certain that variable X is affecting variable Y, and not vice versa. An example would be: X, party voted for at time n; Y, party voted for at time $n + 1$. If we interview thousands of people we will find that most have voted the same way on both occasions. For any individual, the 'value' of X could be used to predict the 'value' of Y. Because of the temporal ordering it makes sense to talk of X affecting Y, but not vice versa. In regression terms we would talk of X being the independent variable and Y being the dependent variable, though we might prefer to use the terms *factors* and *responses*.

Q19   Identify the factors and responses in the four-variable Sussex table.

It has always been considered a good idea in statistics to draw pictures of the data, and sociologists, too, are fond of illustrating the complex relationships between variables using diagrams. A standard technique is to draw arrows connecting variables that interact. If X and Y interact, but it is not clear whether X affects Y or Y affects X, then a double-headed arrow is used. For example, suppose there are three variables X, Y and Z. Suppose that X and Y are both factors and interact with one another, while Z depends on X alone. This would be represented by the path diagram shown in Figure 1.

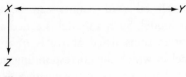

Figure 1
Example of a path diagram

Q20 Draw a path diagram for the data of Table 3.

Path diagrams are used to illustrate straightforward two-variable interactions. Since we are considering several variables simultaneously we may (and indeed already have) come across interactions involving more than two variables. No one has yet hit on a wholly satisfactory way of illustrating a multivariable interaction. This leads to the next questions.

Q21 Suggest a way of illustrating a three-variable interaction.

Q22 Illustrate the interactions present in the four-variable Table 5 using a path diagram.

We noted earlier that some 'interaction totals' were pre-ordained and that therefore we had to restrict attention to models that included the corresponding interaction. Having defined factors and responses we can now be more specific: all models must contain the all-factor interaction. Factors might be described as facts of life: there is no doubt that the English male students interviewed a total of 24 male respondents. All attention focuses on the way that those who were interviewed responded. Thus Question 18 above restricted attention to models including the all-factor BCD interaction.

THE LOG-LINEAR MODEL

The log-linear model deals with linear combinations of (the logarithms of) the original cell frequencies, such transformed statistics being more easily interpretable than the original mass of data. As an example consider a three-variable contingency table in which each variable has two categories. The eight cell frequencies provide us with eight pieces of information concerning this experiment. We can make a transformation of these eight frequencies so that the same information is conveyed by eight quite different statistics as follows:
(a) the overall 'average' frequency (one parameter),
(b) measures of the main effects of A, B and C (three parameters),
(c) measures of the interactions between A and B, A and C, and B and C (three parameters),
(d) a measure of the three-variable interaction (one parameter).

Thus an eight-cell table of counts can be exactly reproduced by an eight-parameter model. Such a model is called a *saturated model*. If one or more of the interactions listed above is of no importance then we can envisage a model in which the corresponding parameter is set to zero. For every parameter that is set to zero we gain a degree of freedom, leading to the following questions.

Q23   In the independence model there are no interactions. How many degrees of freedom does the independence model have for a 2^3 table?

As we set more and more parameters to zero, so we get simpler and simpler models that fit the data less and less well. To test whether a model provides an adequate fit it is necessary to compare the value of a standard goodness-of-fit statistic (e.g. Pearson's statistic) with tables of the chi-squared distribution that has degrees of freedom equal to the number of parameters set to zero.

When the classifying variables have more than two categories the calculations become slightly more complex, as suggested by the following questions.

Q24   Suppose variables X and Y have I and J categories respectively. Given that it requires $(I - 1)$ parameters to account for the variation in frequency between the categories of X, determine the complete breakdown of the IJ parameters in the saturated model for this table. How many degrees of freedom would there be for the independence model fitted to such an $I \times J$ table?

Q25   If variables X, Y and Z have I, J and K categories respectively, draw up a complete table to show how the IJK cell counts are replaced by IJK parameters in the saturated model.

There are now many computer programs that specialize in fitting models to contingency tables. Some programs (e.g. GLIM, SPSS) exploit the parallelism between log-linear models and the more familiar models of regression and the analysis of variance and produce output that is noticeably similar for the two techniques. Other programs (e.g. ECTA, BMDP) use a faster maximum likelihood approach that only deals with the contingency table situation. General advice is difficult, since each program has advantages and disadvantages. Probably the best advice is to use whichever of these programs is most readily available, without worrying whether it is the best.

Q26 Use an available computer program to investigate the data of Table 5 and to find the most appropriate log-linear model. Compare this model with your previous findings.

REFERENCES

Bishop, Y. M. M., Fienberg, S. E., and Holland, P. W. (1975). *Discrete Multivariate Analysis: Theory and Practice*, MIT Press, Cambridge, Massachusetts.

Fienberg, S. E. (1980). *The Analysis of Cross-Classified Categorical Data*, MIT Press, Cambridge, Massachusetts.

Fingleton, B. (1984). *Models of Category Counts*, Cambridge University Press, Cambridge.

Gilbert, G. N. (1981). *Modelling Society*, Allen and Unwin, London.

Sissons, M. (1981). Race, sex and helping behaviour, *British Journal of Social Psychology*, **20**, 285–92.

Upton, G. J. G. (1978). *The Analysis of Cross-tabulated Data*, Wiley, Chichester.

SUPPLEMENTARY QUESTIONS

1. Surveys are often constructed in which one question is answered only if a certain answer has been received to another question. This can result in a cross-classification containing impossible combinations. For example, suppose X is 'vote at time n' and Y is 'vote at time $n + 1$, *if different to vote at time n*'. If the categories of both X and Y are 'party a', 'party b', and so forth, then there will be a leading diagonal of impossible cells. The jargon term is *structural zeros*. How would these affect an analysis of a table?

2. It is evident that when there are many classificatory variables there may be hundreds or thousands of possible alternative models. Suggest some techniques that might be used to avoid the necessity to fit each model in turn. (See Aitkin, 1979, 1980; Brown, 1976.)

3. The hierarchical models that we considered were most appropriate in the case of nominal-level variables. Often one or more of the variables is ordinal, in which case the so-called logit models are more appropriate, since these take account of the ordering of the categories of a variable. How is a logit defined? What is the connection between a log-linear model and a logit model?

4. Consult your local computer software library: there are probably several different computer packages that are available, both on mainframe and on micros, that will fit log-linear models. Contrast their facilities and the time that they take to analyse Table 5.

FURTHER REFERENCES

Aitkin, M. (1979). A simultaneous test procedure for contingency table models, *Applied Statistics*, **28**, 233–42.

Aitkin, M. (1980). A note on the selection of log-linear models, *Biometrics*, **26**, 173–8.

Brown, M. B. (1976). Screening effects in multidimensional contingency tables, *Applied Statistics*, **25**, 37–46.

THE DESIGN AND ANALYSIS OF EXPERIMENTS

- Introduction
- A Bad Experiment in Chemical Process Development
- A Factorial Experiment to Help Determine Government Policy on Bean Growing
- A Split Plot Experiment and a Blocked Experiment to Support the Marketing of Shampoos
- A Hill Climbing Approach to find the Optimum Conditions for Satisfying Quality and Quantity Specifications

Assignments in Applied Statistics
Edited by S. Conrad
© 1989 John Wiley & Sons Ltd

Introduction

Roland Caulcutt

Management Centre, University of Bradford

Why are you interested in the design of experiments and the analysis of experimental results? Perhaps you are studying experimental design as part of a course in statistics. Perhaps you expect that you will, at some time in the future, wish to advise someone on how an experiment should be designed. Perhaps you wish to be able to assess the quality of experiments designed by others.

Whatever your motivation for reading this introduction, it is important that you know what to expect from the four assignments that follow. They will not transform you into an experienced consultant who is able to advise clients on how their experiments should be carried out. To achieve that laudable objective would require that you be confronted with new situations and asked to design experiments which are appropriate. Unfortunately, it is not possible to do this realistically with a written text. In practice, advising on the design of experiments is intimately linked with that most elusive of consultancy skills, the ability to elicit from the client the true nature of the problem. This skill can only be learned from real experience or from a very participative and interactive course.

In the four assignments, therefore, you will not be asked to *design* experiments. You will be required to draw conclusions from experimental results and to comment on the suitability of the design that was used. It will be possible for you to learn a great deal from carrying out the assignments. However, before you do, consider some simple questions about the fundamental nature of experiments. Reflecting upon these questions might help you to see the assignments in a broader perspective.

WHAT IS AN EXPERIMENT?

An experiment is carried out in order to obtain information. 'Isn't this also true of a survey?' you may ask. Yes it is. Surveys and experiments are two types of investigation which differ in one important respect—in an experiment we deliberately change at least one variable. Whereas in a survey we simply observe without interference, in an experiment we introduce change in the hope that we will learn what effect this has.

WHO CARRIES OUT EXPERIMENTS?

It is common knowledge that experiments are carried out by scientists. Management scientists, social scientists, physical scientists and life scientists all carry out experiments both in the laboratory and in the 'field'. Any one of these scientists might turn to the statistician for guidance with the planning of the experiment and/or the analysis of the results.

Perhaps it is not widely realized that *all* humans carry out experiments. You, for example, have probably initiated *many* experiments every day of your life since birth. Psychologists tell us that we continually strive to make sense of our environment by active experimentation. An experiment can provide the means by which we establish linkages between cause and effect. For example, as a baby you may have found that banging your head with your rattle was followed by pain, and concluded that the former was the cause of the latter. Hopefully you made this discovery quite quickly, and then discontinued the experiment. On the other hand, if you made the discovery much later, after you had mastered Latin, you might have exclaimed 'Post hoc ergo propter hoc', thus declaring your faith in the principle that, if Y follows X, it is likely that Y is caused by X. Certainly there is no doubt in most minds that an experiment is likely to tell us more about cause and effect than would a survey.

WHAT IS THE PURPOSE OF AN EXPERIMENT?

We have already observed that scientists, and others, carry out experiments in order to gain a better understanding of their environment. Just as there are many types of environment there are many types of experiment. It is useful to categorize experiments according to the purpose with which they are carried out, namely into:

(a) experiments that are carried out in order to make comparisons,
(b) experiments that are carried out in order to explore relationships between variables.

Comparative experiments are used to compare two or more products, materials, methods, treatments, etc. For example, a manufacturer of shampoos might wish to compare the performance of a new formulation with his existing brand and with several brands made by competitors. An agricultural researcher might wish to compare the effectiveness of several insecticides under certain conditions. A government department might wish to explore the performance of a new analytical test method in various laboratories. The purpose of the experiment is to compare the precision achieved in the different laboratories and to assess the repeatability and reproducibility of the test method.

Experiments in the second category focus upon relationships between variables in order to gain a better understanding of complex systems. For example, a research and development chemist deliberately changes the operating conditions of a chemical plant to determine what effect this variation has upon the quantity of the end product. The increased understanding of the process that results from this experiment will help the chemist to choose 'better' operating conditions.

The two categories are not mutually exclusive. It is possible, in an experiment, to make comparisons under varying conditions. The agricultural researcher, for example, might wish to compare several insecticides using two or more concentrations of each on plants grown with different levels of fertilizer.

WOULD YOU RECOGNIZE A BAD EXPERIMENT?

There are many statistics texts which describe a variety of experimental designs. Readers rightly assume that these designs are *good* in some respects or they would not be recommended. However, many authors do not stress the finer features of the classical designs, nor do they include any *bad* designs with which to contrast them. This is unfortunate because, in practice, the statistician is often presented with the results of bad experiments.

If you were presented with a set of data would you be able to assess how good or bad was the experiment from which the data came? This question is obviously important, but it is not very meaningful unless we define just what we mean by good and bad experiments. We shall define a good experiment as one that yields valid conclusions. In the same style we shall define a good experimental *design* as one that is likely to lead to valid conclusions. On the other hand, a bad experimental design is likely to give misleading results.

In the first of the four assignments you are asked to analyse the results of a bad experiment. The scientist who ordered the experiment to be carried out knows that all is not well because the regression analysis program gives several equations that are contradictory. You are required to advise the scientist on exactly what is wrong with the experiment and what conclusions can safely be drawn from the data. Perhaps this first assignment will not relate directly to any particular lecture or topic within your course. You may, therefore, wish to leave the first assignment until you have done the other three, which are more specific.

WHICH EXPERIMENTAL DESIGNS ARE ACTUALLY USED?

One source of frustration for the statistics student is the sheer multiplicity of techniques presented in many texts. Commonsense suggests that no one uses *all* of these techniques, but the text may give little or no indication as to who uses what. Perhaps some students assume, quite wrongly, that all techniques are equally important, equally useful and equally popular.

The same frustration may arise when studying experimental design. As the student progresses from randomized blocks, to latin squares, to greco-latin squares, to . . . , the mathematical beauty of the designs may increase, but does the usefulness? 'How frequently is each type of experiment actually used?', the student may wonder, but no answer is forthcoming. The reason for this apparent conspiracy of silence could be that no one knows. It is obviously not possible for *any* statistician to have sufficient practical experience in *every* field of application to be able to advise on how frequently each design is used.

Experience in advising scientists, technologists and managers working in research, marketing, production, personnel and distribution, in the chemical and allied industries, is reflected in the four assignments that follow. The second assignment, for example, focuses on 2^n factorial designs and fractional factorial designs. In a manufacturing industry these are used much more than, say, randomized blocks or lattice square designs.

WHAT IS RESIDUAL VARIATION?

An experiment is not an end in itself. The purpose of carrying out experiments is to draw conclusions or to answer questions. Perhaps we wish to know 'Which is the best of the four fertilizers?' or 'What is the effect on yield of changing temperature and feed rate?'. To obtain answers to such questions we would use significance tests and/or confidence limits based on the results of the experiment. Whether we use an F-test, a t-test or confidence limits for a mean, we will require a residual standard deviation or a residual variance.

This may be obtained from an analysis of variance or by other means, but it is essential that this residual should reflect the variability that would have existed if the experimental conditions had not been changed. In other words, the residual standard deviation must be a measure of the variation in the dependent variable that we would have found if the independent variables had been held constant.

To be more specific, suppose that a product evaluation chemist wishes to compare the anti-dandruff efficacy of four shampoos. He randomly assigns his 100 subjects to four groups of 25. Each group will use a different shampoo. The dandruff of each subject is scored by trained assessors before and after a programme of washings. The decreases in dandruff for the 100 subjects are used to compare the four shampoos. Judgements will be based on the four means, but the difference between any two means must be set against the residual standard deviation. This residual is a measure of the variation in decreases in dandruff that we would have found if all subjects had used the *same* shampoo.

Commonsense tells us that the chance of concluding 'shampoo A is better than B' will depend upon the magnitude of the true difference, the size of the residual variation and the number of subjects. To make the experiment more likely to detect a real difference we could increase the number of subjects. This would increase the cost, of course. It might be preferable to give careful thought to reducing the residual variation.

The residual variation arises mainly because subjects achieve differing reductions in dandruff even when they are using the *same* shampoo. (Some of the residual variation will be due to the unreliability of the dandruff assessment, but we will ignore this source of variation.) Would it be possible to reduce the subject to subject variation? Alternatively, would it be possible to redesign the experiment so as to eliminate the subject-to-subject variation altogether? The latter would be achieved if each subject used all four shampoos.

In the language of the statistician, applying all four shampoos to each head would be described as *blocking*. This is a technique much used in agricultural experiments, when the available land is split into small, and thus homogeneous, blocks with all treatments being applied to every block. The advantages and disadvantages of blocking feature prominently in the third assignment.

A LARGE EXPERIMENT OR SEVERAL SMALL ONES?

All experimenters must work within constraints when designing their experiments. Perhaps the most obvious constraint in agricultural experiments is the need to respect the growing season of the plants. Because of this, many agricultural experiments take a year or more to plan, execute and report. Should further experimentation be required this would take a second year. Thus the agricultural researcher is strongly motivated to 'get it right first time' and to answer *all* research questions with *one* experiment.

Few other researchers suffer this severe time constraint. For example, many industrial experiments can be carried out in one month, one week, or even one day. Thus the industrial scientist may be more concerned with minimizing cost or disturbance than with the time that is needed to wait for results. Clearly, it is not wise to put all your eggs in one basket, and the industrial statistician may well advise a client to use a series of small experiments, with the planning of the second being deferred until the results of the first have been analysed. It is often suggested that no more than 30 per cent of the experimental budget should be allocated to the first experiment.

This philosophy of sequential experimentation is at the heart of the response surface methods in the last of the four assignments. A 'hill climbing' approach is used to progress towards the operating conditions that will give the best performance of an industrial process. It is quite common to carry out a succession of fractional factorial experiments when using these methods.

Assignments in Applied Statistics
Edited by S. Conrad
© 1989 John Wiley & Sons Ltd

A Bad Experiment in Chemical Process Development

Roland Caulcutt

Management Centre, University of Bradford

OUTLINE

It is quite easy to analyse data from an experiment that has followed a classical design such as a latin square, a 2^n factorial or a split plot. Unfortunately many scientists and technologists know little or nothing about these designs, and they often bring to the statistician data which are much more difficult to analyse. This assignment focuses upon such a set of data. Perhaps the greatest benefit to be gained from studying a 'bad' experiment is that it helps you to appreciate the fine qualities of the classical designs, which are so often taken for granted. The use of multiple regression analysis with this bad set of data illustrates the power of this technique but also highlights the dangers associated with its use.

KEYWORDS

simple regression analysis, multiple regression analysis, residuals, percentage fit, computer-aided design of experiments, intercorrelation of independent variables, interaction between independent variables, communication between statistician and client

Dr Mawgan is a chemical engineer in the Research and Development Department of Colour Chemicals plc. He is responsible for process investigations which are carried out in the hope of discovering ways of improving production processes. In these investigations experiments are carried out, either on a small-scale pilot plant or on the full-scale production plant.

One particular investigation was initiated by Dr Mawgan, following complaints from customers about the quality of 'digozo blue'. This dyestuff is widely used in the textile industry but it is also manufactured by three competitors. It is, therefore, important that the cause of the reduced quality should be determined and corrective action taken.

Discussions with a textile technologist led Dr Mawgan to the conclusion that the poor performance of the dyestuff is due to excessive concentrations of a particular impurity. He wonders how this impurity can be reduced. With a complex chemical manufacturing process we can only control the nature of the product indirectly, by changing the operating conditions. Thus Dr Mawgan needs to know how the concentration of the impurity is dependent upon those independent variables which can be manipulated. Each operator, each manager and each scientist has opinions on how the process should be controlled. However, there is little agreement between them. Clearly an experiment is required in order to obtain a better understanding.

It is decided that this experiment will be carried out on the full-scale production plant and that five variables will be explored. Ten batches of digozo blue will be produced using different levels of these variables. Dr Mawgan hopes that the resulting impurity in the ten batches will help him to answer five questions:

1. What weight of triazone should be used in each batch in order to reduce the impurity to less than 4 per cent? It is known without doubt that the addition of triazone does reduce impurity. However, this ingredient is very expensive and Dr Mawgan would be pleased if he could demonstrate that a satisfactory level of impurity could be achieved without the inclusion of triazone.

2. How is the impurity in the digozo blue dependent upon the feed rate and the temperature of the main ingredient? Both of these variables can be controlled very easily within certain limits. Increasing the temperature and/or reducing the feedrate would increase the cost of each batch but the extra cost would be negligible.

3. What is the best speed at which to run the mechanical agitator? It is important that the mixture should be stirred during the chemical reaction but it can be predicted on theoretical grounds that too high or too low a speed might result in excessive impurity.

4. Does impurity increase as the catalyst ages? An expensive platinum

catalyst is used to speed up the chemical reaction. After several batches the catalyst becomes contaminated and the yield decreases (i.e. from the same input we obtain less digozo blue pigment). Thus it has been the practice to change the catalyst after approximately twenty batches. However, Dr Mawgan suspects that the impurity may increase as the catalyst deteriorates and he wonders if it would be wise to change the catalyst earlier.

5. If the impurity *does* increase as the catalyst ages, would it be possible to compensate for this by changing some other variable? Perhaps increasing the agitation speed or decreasing the feedrate would reduce the impurity in later batches to the acceptable level in those batches produced soon after a catalyst change.

(Note: you are not expected to answer the above questions. They are presented here so that you will have no doubt about the purpose of Dr Mawgan's experiment.)

Having decided which variables will be included in this experiment Dr Mawgan now intends to make use of a computer program to design his experiment. He realizes that it might be possible to obtain a suitable design from a book, but Dr Mawgan cannot face the painful task of searching through a statistics text, and he certainly does not wish to face the company statistician. Past experience has taught Dr Mawgan that the statistician prefers to change the problem until it fits a classic experimental design, rather than seek a solution to the existing problem.

The computer program needs to know what values will be used for each independent variable. Dr Mawgan specifies that 'weight of triazone' will have six values: 0.5, 1.0, 1.5, 2.0, 2.5 and 3.0 kg. He hopes that using a large number of different weights will help him to determine the effect of this variable with great accuracy. For each of the other independent variables he specifies only three values. He asks for 'feed rates' of 35, 40 and 35 litres per minute, 'temperatures' of 80, 85 and 90°C, 'agitation speeds' of 10, 20 and 30 revolutions per minute. The computer program gives him the design in Table 1.

Dr Mawgan does not understand how the program operates, even though he has read the manual which accompanied the program. This manual explained the principles on which the program was based, but it was written in the language of the statistician. Thus, Dr Mawgan could not explain *why* the design in Table 1 is a good design. Indeed, if the computer had given him *two* experimental designs, it is doubtful if he could have selected the better, with any confidence. Nonetheless, Dr Mawgan is well aware that a *good* design is likely to lead to unambiguous conclusions, whereas a *bad design* might leave him in great doubt about the operating conditions required to achieve low levels of impurity. (See Caulcutt, 1983, Ch. 11.)

Table 1
A computer generated experimental design

WEIGHT OF TRIAZONE	FEEDRATE	TEMPERATURE	AGITATION SPEED
x	z	t	s
1.0	40	80	20
3.0	45	80	10
2.5	35	85	30
2.0	45	90	30
2.5	40	90	20
1.5	45	85	20
3.0	35	85	20
1.0	45	85	10
0.5	40	80	30
0.5	35	90	10

Dr Mawgan discusses Table 1 with the Plant Manager, without whose cooperation no experiment could take place. The Plant Manager agrees to produce ten batches using the specified operating conditions and suggests that the experiment should start immediately after the next catalyst change. Dr Mawgan will be on holiday at the time but he is confident that his instructions will be followed with care. On his return he is presented with the results of the experiments which are summarized in Table 2.

Dr Mawgan studies Table 2 very carefully. He is very relieved to find that in only one of the ten batches was the impurity intolerably high. Furthermore, he is very pleased that three batches (684, 688 and 691) had such low impurity. Before putting the data into his computer for analysis

Table 2
The actual experiment

BATCH NUMBERS	WEIGHT OF TRIAZONE	FEEDRATE	TEMPERA- TURE	AGITATION SPEED	CATALYST AGE	IMPURITY
	x	z	t	s	w	y
683	2.0	38	80	20	1	4.2
684	3.0	45	80	10	2	2.7
685	1.0	35	85	30	3	6.6
686	2.0	45	90	30	4	5.9
687	3.5	50	90	20	5	2.2
688	2.5	47	85	20	6	3.1
689	1.0	36	85	20	7	6.3
690	3.0	47	85	10	8	5.2
691	1.5	39	80	30	9	2.9
692	0.5	30	90	10	10	7.5

he plots five graphs to illustrate the relationship between impurity and each of the five independent variables. Whilst plotting these graphs he realizes that the operating conditions used during the experiment were not exactly what he had specified. A comparison of Tables 1 and 2 reveals that the temperature and the agitation speed were controlled exactly as he had requested. However, the feed rate and the weight of triazone were, in several batches, rather different from the values specified in Table 1. This is most noticeable with batch 687.

There is no question of repeating the experiment. Dr Mawgan must attempt to draw conclusions from the data in Table 2, but before doing so he interviews the staff who were involved to ensure that Table 2 truly represents what actually happened. He is assured that the operating conditions listed in Table 2 are correct.

To analyse the data he makes use of a multiple regression program on his personal computer. This gives him the following equations:

$$\text{IMP} = 7.25 - 1.43 \ \text{TRIAZONE} \qquad 57.4\% \ \text{fit}$$

$$\text{IMP} = 13.1 - 0.204 \ \text{FEEDRATE} \qquad 49.2\% \ \text{fit}$$

$$\text{IMP} = -11.8 + 0.193 \ \text{TEMP} \qquad 17.5\% \ \text{fit}$$

$$\text{IMP} = 4.66 - 0.0000 \ \text{SPEED} \qquad 0.0\% \ \text{fit}$$

$$\text{IMP} = 3.63 + 0.257 \ \text{CAT. AGE} \qquad 13.5\% \ \text{fit}$$

A significance test on the percentage fits (see Caulcutt, 1983, pp. 139 and 156) reveals that only the first two equations are statistically significant at the 5 per cent level. Dr Mawgan concludes that impurity will be reduced by 1.43 per cent for every additional kilogram of triazone added. Though triazone is very expensive such a large reduction in impurity is well worth the extra cost. He further concludes that an increase in feedrate of 10 litres per minute will give a reduction in impurity of 2.04 per cent. This second conclusion is even more interesting as the feedrate can be increased with little additional cost and can probably be increased beyond 50 litres per minute, which was the highest value used in the experiment. The prospect of reducing the impurity by increasing *both* feedrate *and* weight of triazone is very exciting indeed.

In this report, Dr Mawgan must specify what weight of triazone and what feedrate should be used to achieve an acceptable level of impurity. In order to obtain the predicted impurity for different values of the two independent variables he asks the program to fit a multiple regression equation. It gives him:

$$\text{IMP} = 6.71 - 1.60 \ \text{TRIAZONE} + 0.028 \ \text{FEEDRATE} \quad 57.5\% \ \text{fit}$$

As Dr Mawgan substitutes values of feedrate and triazone into this equation to obtain predicted impurity, he becomes very concerned on two counts:

1. The percentage fit is only 57.5 per cent, which is little better than the 57.4 and 49.2 per cent obtained with the two simple equations considered earlier. He had expected a much higher percentage fit, though he realized that 106.6 per cent (i.e. 57.4 plus 49.2) was not possible.
2. The feedrate coefficients in the two equations are contradictory. The simple equation implies that feedrate should be *increased* in order to reduce impurity whereas the multiple equation suggests that a *decrease* in feedrate is required.

Dr Mawgan does not know why his data yield these ambiguous conclusions. He suspects, however, that he would be in a much better position if the Plant Manager had carried out the experiment specified in Table 1. He decides to ask for the help of the company statistician.

THE ASSIGNMENT

You are the Company Statistician. You joined Colour Chemicals plc two years ago, immediately after gaining your MSc in statistics from Brudfax University. Your career development plan included a lengthy induction period under the watchful eye of Dr Dartah who had been the Company Statistician for many years. Unfortunately he was involved in an accident on the day you arrived and never returned to work.

In your early days with Colour Chemicals you were inundated with requests for assistance from scientists and managers in several departments. Many of these requests seemed to have little statistical content and many of the clients appeared to assume that you had great technical knowledge of their particular situation. You had great difficulty coping with the innumerable demands made upon you in your first weeks with the company. Gradually, however, the pressure eased and recently you have had very few callers.

When Dr Mawgan enters your office you realize that he is the first client who has paid you a second visit! You are now, of course, much more experienced as a statistical consultant than you were when he first called upon your services. You now realize the paramount importance of listening carefully to the client's description of his problem and attempting to communicate with him in his own 'language'. See Boen (1972), Greenfield (1979), and Marquardt (1979), for a discussion of the issues surrounding and difficulties experienced by statisticians communicating with clients.

Dr Mawgan explains the background to the problem and gives you a list of the five questions he posed at the outset. He shows you Table 1 and Table 2 and his regression equations. He appeals to you to rescue him from his immediate predicament and to help him avoid any recurrence. In order to help Dr Mawgan, you are asked to answer the following questions.

Q1   Use a multiple regression program to fit equations of the type fitted by Dr Mawgan. Are his equations correct?

Q2   The reason why Dr Mawgan's multiple regression equation:

$$IMP = 6.71 - 1.60 \ TRIAZONE + 0.028 \ FEEDRATE$$

contradicts his simple regression equations:

$$IMP = 7.25 - 1.43 \ TRIAZONE$$

and

$$IMP = 13.1 - 0.204 \ FEEDRATE$$

can be found in Table 2. Carry out any analysis of the data in Table 2 that is necessary to identify the major defect in the experiment. (See Chatterjee and Price, 1977; Belsey, Kuh and Welsch, 1980, Ch. 3; Caulcutt, 1983, Ch. 11; Weisberg, 1985, Ch. 8.)

Explain to Dr Mawgan, in terms that he would understand, why the results of his experiment do not lead to clear unambiguous conclusions.

Q3   Before the experiment was carried out Dr Mawgan suspected that impurity increased as the catalyst aged. Use multiple regression analysis to explore this possibility and explain your findings to him.

Q4   Dr Mawgan also suspected that the effect on impurity of varying the weight of triazone, the feedrate, the temperature or the agitation of speed, might change as the catalyst aged. To examine this possibility plot four scatter diagrams with weight of triazone, feedrate, temperature and agitation speed on the vertical axis, using catalyst age on the horizontal axis in each case. On all four diagrams write the value of impurity next to each point. What conclusions can you draw from these diagrams?

Q5   The diagrams you drew in Question 4 are very simple, but such diagrams can be very revealing. Unfortunately they are tedious to draw by hand and few computer programs are capable of labelling points on a scatter diagram. How would you use a multiple regression program to examine numerically the relationships that you explored graphically in Question 4? (See Chatterjee and Price, 1977, p. 78; Caulcutt, 1983, pp. 162, 184 and 228; Weisberg, 1985, p. 165.)

Q6 Use your multiple regression program to seek significant interactions between the five independent variables. What conclusions can you draw concerning the best operating conditions for the production process? How do your conclusions compare with those drawn in Question 4?

Q7 If further experimentation were not possible, decisions concerning the best operating conditions would have to be based on:

(a) the data in Table 2,
(b) prior knowledge of the process.

Unfortunately, you have very little prior knowledge. Specify operating conditions that you consider will give a mean impurity in future batches of less than 3 per cent.

Q8 We have seen that the experiment that was carried out (Table 2) was not a good experiment. Perhaps the intended experiment (Table 1) was much better. Add an extra column for catalyst age then critically examine Table 1. Would you recommend any changes?

Q9 Dr Mawgan has persuaded the Plant Manager to produce a further eight batches of digozo blue under experimental conditions. Specify operating conditions for these eight batches so that all eighteen batches together will constitute a 'good' experiment which is likely to yield clear conclusions. (You will wish to choose values for the five independent variables so that the correlation matrix for all eighteen batches does not contain the collinearity that you found in Table 2.)

Q10 Suppose that the data from all eighteen batches is analysed using stepwise multiple regression analysis and the 'best' equation gives the residuals in Figure 1. (The numbers in the circles indicate the order in which the batches were produced.) What conclusions can you draw from these residuals? How could the regression analysis be extended to account for the non-randomness in Figure 1?

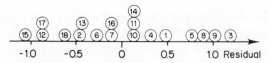

Figure 1
Residuals from multiple regression

REFERENCES

Belsey, D. A., Kuh, E., and Welsch, R. E. (1980). *Regression Diagnostics*, Wiley, New York.

Boen, J. (1972). The teaching of interpersonal relationships in statistical consulting, *The American Statistician*, 26(1), 30–1.

Caulcutt, R. (1983). *Statistics in Research and Development*, Chapman and Hall, London.

Chatterjee, S., and Price, B. (1977). *Regression Analysis by Example*, Wiley, New York.

Greenfield, A. A. (1979). Statisticians in industrial research: the role and training of an industrial consultant, *The Statistician*, 28, 71–82.

Marquardt, D. W. (1979). Statistical consulting in industry, *The American Statistician*, 33, 102–7.

Weisberg, S. (1985). *Applied Linear Regression*, Wiley, New York.

SUPPLEMENTARY QUESTIONS

1. In this assignment reference has been made to 'computer aided design of experiments'. What are the principles underlying the algorithms that generate these designs? (See Kennard and Stone, 1969.)

2. Stepwise multiple regression usually incorporates a sequence of significance tests. What assumptions underly these tests? What steps can be taken to overcome any violation of these assumptions?

3. In this assignment some importance has been attached to interactions between pairs of independent variables. Hopefully the reader will realize that interactions are not rare phenomena, but common features of everyday life. List some of the variables that can be manipulated during the making of a cup of tea. Which pairs of the variables in your list interact in their effect on the quality of the final product?

FURTHER REFERENCE

Kennard, R. W., and Stone, L. A. (1969). Computer aided design of experiments, *Technometrics*, 11, 137–48.

Assignments in Applied Statistics
Edited by S. Conrad
© 1989 John Wiley & Sons Ltd

A Factorial Experiment to Help Determine Government Policy on Bean Growing

Roland Caulcutt

Management Centre, University of Bradford

OUTLINE

The earliest books on experimental design discussed those designs that had proved useful in agricultural research, such as randomized blocks, incomplete blocks or latin squares. However, these designs are little used in industry where the factorial and fractional factorial designs are more popular. In this assignment we focus on a factorial experiment which was designed by an industrial scientist for use in an agricultural study. Unfortunately, he returned to industry and the data analysis was left in the hands of someone who clearly lacked the necessary skills. You are asked to re-analyse the data. You will need some knowledge of 2^n factorial experiments and three-way analysis of variance in the first part of the assignment: you will need some understanding of fractional factorial experiments in the second. A familiarity with normal probability plots will also be useful.

KEYWORDS

factorial experiment, fractional factorial experiment, replication, analysis of variance, assumptions, t-tests, F-test, half-normal plot

Dr Romero is a Principal Administrative Officer employed by the National Agricultural Board of Parazil. He is leading a major research study into the growing of demingo beans, which are an important source of protein for the poorer people in Parazil and in neighbouring countries. A prime objective of Dr Romero's study is to improve the yield of the beans, so that a smaller area of land is required for their cultivation, and to improve the quality of the beans which should result in increased exports.

The agricultural scientists in Romero's team have carried out an experiment in which the demingo beans were planted in two types of soil, red and grey. Visitors to Parazil are often struck by the dramatic changes in the colour of the soil as they pass from region to region. The demingo beans have been grown successfully in many regions and on both soil types, but the scientists are hoping to demonstrate that the red soil is more suitable for beans, so that certain regions with grey soil can be turned over to the growing of tobacco.

A second purpose of the experiment was to establish whether it is better to plant the beans in May or in September. Planting in May allows the beans to establish root growth during the winter and take off more quickly in spring for an early harvest in November. (Remember that Parazil is in the southern hemisphere.) September planting gives a later harvest in February which is dangerously close to the wet season. A third purpose of the experiment was to determine the effect on the beans of prolonged storage. If only one planting time is used it will be necessary to store the beans for up to 12 months. Dr Romero has been advised that the quality of the beans will deteriorate in storage and this will be accompanied by a decrease in the chewiness of the beans. Thus the experimenters have taken great care in assessing the chewiness and other sensory properties of the beans, as well as the yield (see Table 1).

Table 1
Chewiness scores

	PLANTING DATE			
	MAY		SEPTEMBER	
	SOIL TYPE		SOIL TYPE	
STORAGE	GREY	RED	GREY	RED
Seven months	4 3 3 3 4	4 3 3 4 3	3 5 2 3 4	3 3 2 1 3
	5 3 1 3 2	2 3 6 4 4	3 6 2 5 4	4 3 4 3 3
One month	3 2 2 2 3	4 6 5 4 7	4 1 4 4 6	4 5 6 4 5
	2 4 3 2 2	4 5 5 6 6	5 4 3 4 4	5 7 4 4 7

The chewiness of the cooked beans was assessed by one trained assessor on a scale from 1 to 9. Obviously this assessment is very subjective and may be influenced by day-to-day variation since the scoring of the 80 samples was spread over several days in each of four sessions. The samples were assessed in random order as far as possible. A high chewiness score is indicative of high quality and it is expected that chewiness scores would be lower for the beans which had been stored longer.

The experiment was designed by an agricultural scientist who was on secondment from the Surface Chemical Company. He has since returned to his employer leaving the project team deficient in statistical expertise. Data analysis is therefore entrusted to a recently graduated biologist whose training in statistics was not continued beyond the t-test. His report contains the following conclusions and recommendations:

> The mean chewiness after 7 months storage was 3.325 compared with a mean of 4.175 after 1 month storage. This difference is significant at the 1% level. The mean chewiness of the beans grown on the red soil was 4.2 whilst the mean for beans grown on the grey soil was only 3.3. This difference is significant at the 1% level. The mean chewiness of the beans planted in September was 3.9 whilst the mean for beans planted in May was 3.6. This difference is not significant even at the 10% level. It is recommended that the demingo beans should be planted on the red soil rather than the grey. It is recommended that planting should take place in May and September to spread the harvesting of the beans over as long a period as possible and thus reducing the time spent in storage.

Dr Romero reads the report with great interest but he finds the conclusions rather disappointing. He had great faith in the industrial scientist who designed the experiment and he had expected that the data analysis would reveal very subtle points that would not be obvious to the untrained eye. The conclusions in the biologist's report are no better than he himself would have drawn after a short visual inspection of the data. Dr Romero had hoped to learn whether the beans grown on red soil deteriorated in storage more slowly, or more quickly, than those grown on grey soil. He had also hoped to learn whether or not the deterioration in storage was the same for beans planted in May and in September. Above all he had expected to establish once and for all what was the best planting time for each of the two soil types. He has been told that farmers in the red soil regions prefer to plant in May whilst farmers on grey soil prefer to plant in September. 'Do the results of the experiment offer any support for this practice?' he wonders. To focus his mind on this particular question he calculates the means in Table 2.

Each mean in Table 2 is the average of twenty chewiness scores from Table 1. Dr Romero can understand a table of means much better than a

Table 2
Mean chewiness with different soils and planting times

PLANTING	SOIL TYPE	
	GREY	RED
May	2.8	4.4
September	3.8	4.0

t-test and he feels that a clear conclusion can be drawn from Table 2:

> If demingo beans are to be planted in grey soil they should be planted in September, but on red soil the beans should be planted in May.

He wonders what type of statistical analysis would be required to draw this obvious conclusion.

Q1 Comment briefly on the statistical analysis carried out by the biologist: in particular,

(a) What are the assumptions underlying the t-test. Are they likely to be violated in this situation? (See Box, Hunter and Hunter, 1978, Ch. 3; Caulcutt, 1983, Ch. 7.)

(b) What other method of analysis would be more appropriate than the t-test and what advantages would it offer? (See Chatfield, 1978, p. 258; Davies, 1978, p. 252.)

Q2 Carry out a three-way analysis of variance on the data in Table 1. (See Davies, 1978, p. 280; Caulcutt, 1983, p. 191.)

Q3 Carry out F-tests to check the statistical significance of the three main effects and the four interactions.

Q4 Draw conclusions and make recommendations. These should be expressed in language which would be readily understood by Dr Romero. Your recommendations should clearly indicate when the demingo beans should be planted, in which type of soil and how much they are likely to deteriorate in storage. If you do not appreciate the necessity for communicating in the language of the scientist consult Joiner (1982) and Nelder (1986).

FURTHER INVESTIGATION

Take another look at the data in Table 1. Each of the eight cells in the table contains ten chewiness scores. The structure of the table implies that:

(a) beans were grown under eight different experimental conditions and

(b) the ten scores in any cell were obtained from ten samples grown under identical conditions.

Whilst carrying out your data analysis you probably assumed that the above was true. Dr Romero made the same assumption. However, when he discusses the experiment with the agricultural scientist who drew up the original plan he discovers that this assumption may not be valid. It appears that the original plan had included five independent variables. In addition to the three variables shown in Table 1 were two others, 'the time of application of potash' and 'the quantity of herbicide'.

Dr Romero questions members of the research team and learns that the experiment was indeed carried out according to the original plan. However, two of the independent variables were ignored during the data analysis because the computer program could not cope with them. Further investigation reveals that, in each cell of Table 1, the five scores in the upper row are from beans on which the potash was applied early, whilst the five scores in the lower row are from beans on which the potash was applied later. Furthermore, in four of the eight cells only 10 grams per square metre of herbicide had been used whilst in the other four cells this was increased to 20 grams per square metre. The full details of the designs are set out in Table 3.

Dr Romero is very disturbed. It appears that the experiment was well designed and carefully executed, but the efforts of many people were then nullified by the incompetence of the data analyst. Romero himself has no statistical training but he feels sure that it is very unwise to deliberately vary five variables and then simply ignore two of them.

It is now clear to Dr Romero that the project team became very deficient in statistical expertise upon the departure of the agricultural scientist who

Table 3
The plan that was followed

	PLANTING DATE			
	MAY		SEPTEMBER	
	SOIL TYPE		SOIL TYPE	
STORAGE	GREY	RED	GREY	RED
Seven months	10g Early	20g Early	20g Early	10g Early
	10g Late	20g Late	20g Late	10g Late
One month	20g Early	10g Early	10g Early	20g Early
	20g Late	10g Late	10g Late	20g Late

planned the experiment. Perhaps it would have been wise to call in a statistical consultant at that time, but the project manager did not do so. Now, with the Minister for Agricultural Development calling for a report which is overdue. Dr Romero must take personal control and seek advice from an external source. He turns to you for guidance, and for answers to the following questions.

Q5   The layout of Table 1 gives the impression that a 2^3 factorial experiment was carried out with ten replicates in each of the eight cells. In the light of the further information now available this is clearly incorrect. Table 3 contains sixteen cells and embraces five independent variables. How would you describe the experiment, now that you have learned of the existence of the other two independent variables?

Q6   As the experiment is not a complete factorial experiment, it will not be possible to estimate all of the five main effects and 26 interactions. List the alias groups to show what can and can not be estimated. (There are many ways in which this can be done. One simple method is to draw up a design matrix as in Davies, 1978, p. 460; Caulcutt, 1983, p. 201.)

Q7   Use one-way analysis of variance to break down the total variation in Table 1 into two components:
(a) within the sixteen cells.
(b) between the sixteen cells.

Q8   The 'between cells sum of squares' calculated in Question 7 can be broken down into fifteen components using Yates' method or the design matrix or other techniques. Carry out this further subdivision and produce an analysis of variance table. (See Johnson and Leone, 1964, p. 182; Davies, 1978, p. 264.)

Q9   As a result of your analysis of variance what conclusions can you draw concerning the effect on chewiness of the five independent variables?

Q10   What assumptions underly the analysis of variance carried out in Question 2? How can these assumptions be checked? (See Johnson and Leone, 1964, pp. 4 and 24; Barnett and Lewis, 1978, Chs. 1 and 2; Caulcutt and Boddy, 1983, p. 156.)

Q11   Each of the sixteen cells in Table 3 contained five replicates. The residual mean square in the analysis of variance is based on the variation within cells. Thus without the replication there would be no residual. However, if we are prepared to make additional assumptions, we can obtain a residual mean square from a single replicate of a factorial experiment. One approach to this problem was suggested by Daniel (1959). This involves plotting the effect estimates on normal probability paper or on 'half-normal' paper. Plot the fifteen effect estimates from Question 8, either on half-normal paper or on the more readily obtainable normal probability paper. What conclusions can you draw from your plot? (See Daniel, 1976; Box, Hunter and Hunter, 1978.)

REFERENCES

Barnett, V., and Lewis, T. (1978). *Outliers in Statistical Data*, Wiley, Chichester.

Box, G. E. P., Hunter, W. G., and Hunter, J. S. (1978). *Statistics for Experimenters*, Wiley, New York.

Caulcutt, R. (1983). *Statistics in Research and Development*, Chapman and Hall, London.

Caulcutt, R., and Boddy, R. (1983). *Statistics for Analytical Chemists*, Chapman and Hall, London.

Chatfield, C. (1978). *Statistics for Technology*, Chapman and Hall, London.

Daniel, C. (1959). Use of half-normal plots in interpreting factorial two-level experiments, *Technometrics*, **1**, 311–41.

Daniel, C. (1976). *Applications of Statistics in Industrial Experimentation*, Wiley, New York.

Davies, O. L. (1978). *The Design and Analysis of Industrial Experiments*, Longman, London.

Johnson, N. L., and Leone, F. C. (1964). *Statistics and Experimental Design, Volume 2*, Wiley, New York.

Joiner, B. L. (1982). 'Consulting, Statistical'. In S. Kotz and N. L. Johnson (eds.), *Encyclopedia of Statistical Sciences, Volume 2*, Wiley, New York, pp. 147–55.

Nelder, J. A. (1986). Statistics, science and technology, *Journal of the Royal Statistical Society (Series A)*, **149**, 109–21.

SUPPLEMENTARY QUESTIONS

1. Analysis of variance, because it is so widely applicable, helps to unify the discipline of statistics. Not surprisingly it is much loved by statisticians. However, in most applications of analysis of variance there will be alternative methods of analysis, and often these will be simpler. The alternative methods would probably appeal more to the client, if not to the statistician. Comment on the rather extreme view that 'in no situation is analysis of variance the best technique to use'.

2. Parametric techniques such as t-tests and F-tests are often used in practice, even when it is doubtful that their assumptions are satisfied. Which non-parametric techniques could be used to analyse the data from the demingo bean experiment?

3. Some data analysts have great faith in multiple regression analysis. Perhaps they sometimes use this very powerful technique when analysis of variance would be more appropriate. How would you set up the five independent variables in order to analyse the bean data by multiple regression analysis?

Assignments in Applied Statistics
Edited by S. Conrad
© 1989 John Wiley & Sons Ltd

A Split Plot Experiment and a Blocked Experiment to Support the Marketing of Shampoos

Roland Caulcutt

Management Centre, University of Bradford

OUTLINE

When advising a client on the planning of an experiment the statistician may be tempted to change the objectives in order to use a simple classical design. In many situations, however, it is desirable to remain faithful to the original objectives even if this necessitates the use of a much more complex or non-standard design. Of course, the statistician can give a balanced opinion only if fully aware of the advantages and disadvantages of each. In this assignment we consider a split-plot design used to compare the anti-dandruff properties of two shampoos. The disadvantage of the split-plot design will not need to be emphasized as you struggle to extract the two residual mean squares, but the advantage of using this design should also emerge very clearly. You will realize that the advantage is only gained if the treatment combinations are blocked in a suitable way. In the second part of the assignment you are asked to consider other aspects of blocking in connection with an experiment designed to compare alternative video presentations.

KEYWORDS

blocking, split plot design, within blocks residual, between blocks residual, F-test, interaction, confounding, defining contrast, alias pairs

Gordon Staples is a Marketing Manager with Beauticare Limited. He has overall responsibility for hair care products, including the development of new brands and the marketing of existing brands. During the five years Gordon has been with Beauticare and the previous eight years spent with Duocurl, he has acquired considerable knowledge of the shampoo market. He is well aware, for example, that Beauticare's best selling shampoo. Cleartop, has recently been displaced as market leader by Duocurl's product, Softsheen. He is also aware that his position within the company would be greatly improved if he could reverse this relative decline.

The marketing of shampoos is very competitive, of course, with vast expenditure on television and magazing advertising. Undoubtedly a successful advertising campaign would re-establish Cleartop but the most effective TV advertisements are those that claim that the product is 'best' and Gordon has no conclusive evidence that Cleartop is better than Softsheen in any important respect. However, he is hopeful that the research department will soon produce such evidence. (You realise, no doubt, that the Advertising Standards Authority continuously monitors all press and television advertising. The ASA can, and frequently does, demand to see evidence in support of claims made in advertisements. The failure to produce satisfactory evidence often results in adverse publicity and the imposition of financial penalties.)

During the past three years two experiments have been carried out at Beauticare with the specific purpose of comparing the two shampoos, Cleartop and Softsheen. For both experiments male members of the public were recruited by means of advertisements in the local evening paper. Equal numbers of subjects were then assigned at random to each of the two shampoos.

The first experiment focused on the anti-dandruff properties of the two products and the dandruff of each subject was assessed before and after the three-week treatment period. The dandruff scores were awarded by trained assessors using a scale from 1 to 7. The subjects were instructed to wash their hair twice a week during the three-week period, using the shampoo provided, which was in an attractive but unlabelled bottle. Unfortunately 3 of the 25 subjects using Cleartop did not report for the second assessment, nor did 2 of the 25 subjects using Softsheen. The mean decrease in dandruff score for the 22 subjects completing the Cleartop treatment was 2.15 whilst the mean decrease for the 23 Softsheen subjects was only 0.76. This would indicate that Cleartop was more effective as an anti-dandruff shampoo. However, with standard deviations of 2.074 and 1.852 this difference did not prove significant in a t-test. Thus Beauticare were unable to use the results of this experiment to support an advertising claim that 'Cleartop is more effective against dandruff than other leading brands.'

The second experiment attempted to investigate the effect of two

Table 1
Decrease in dandruff: second experiment

	TYPE OF HAIR															
	DRY								GREASY							
	LENGTH OF HAIR								LENGTH OF HAIR							
SHAMPOO	SHORT				LONG				SHORT				LONG			
Cleartop	2	2	−1	3	1	1	3	2	−1	3	3	4	4	3	1	3
Softsheen	0	0	3	1	4	2	−1	3	2	0	−3	1	−1	2	0	1

additional factors which, it was thought, might influence the anti-dandruff properties of the two shampoos. These factors were the dryness/greasiness of the hair and the length of the hair. The inclusion of these additional factors made the screening of potential subjects much more complex as it was now essential to obtain equal numbers of subjects in each of the four categories: dry–short, dry–long, greasy–short and greasy–long. Furthermore, it was desirable to balance the ages of the subjects within each of the four groups: thus many applicants were rejected before the final complement of eight subjects was obtained for each group. Finally four subjects were randomly assigned to each shampoo within each of the four groups. As in the first experiment a dandruff score was recorded for each subject before and after the three-week treatment period. The decrease in dandruff score is given in Table 1 for each of the 32 participants.

When Gordon Staples saw the results of this second experiment he soon reached the conclusion that Cleartop had achieved a greater improvement than Softsheen. After more detailed scrutiny of Table 1 he reached the further conclusion that Cleartop had been even more effective on greasy hair than on dry hair. Gordon did not carry out significance tests. He simply compared each Cleartop result with the corresponding Softsheen result; then he compared each greasy result with the corresponding dry result. (Perhaps he did not realize that, within each of the eight cells, the results are arranged arbitrarily. A different arrangement within the groups might have led him to a different conclusion.)

In due course the data in Table 1 was analysed by the researchers who had planned and carried out the experiment. Three-way analysis of variance was used.

Gordon Staples could see in Table 2 that the mean square for shampoos and the mean square for the interaction between shampoos and hair types are both much larger than the residual mean square. Thus the analysis of variance clearly supported his conclusions. Unfortunately, F-tests on these mean squares showed that they were not significant at the 5 per cent level.

Table 2
Analysis of the data in Table 1

SOURCE OF VARIATION	SS	DF	MS
Shampoos	11.28	1	11.28
Types of hair	0.28	1	0.28
Lengths of hair	2.53	1	2.53
Interaction S × T	9.03	1	9.03
Interaction S × L	0.28	1	0.28
Interaction T × L	0.03	1	0.03
Interaction S × T × L	0.28	1	0.28
Residual	70.25	24	2.93
Total	93.96	31	—

Thus the evidence furnished by the second experiment was also inconclusive and could not be used to support an advertising claim.

At this point Gordon was undecided whether to carry out a further experiment or to abandon the programme. There seemed to be no certainty that it would ever produce a usable conclusion. The cost of the two experiments had taken from his budget money that could have been well spent on promotion of the product. On the other hand, both experiments had given some indication that Cleartop was a superior product. Perhaps a third experiment would yield the required proof.

Before agreeing to further expenditure Gordon Staples wishes to know how the third experiment will improve upon the second. To this end he calls in a statistical consultant who is asked to comment on the work already carried out and to make recommendations for future experimentation. During the discussion the consultant points out that a larger experiment is more likely to produce a significant difference than a smaller experiment. Thus a repeat of the second experiment using, say, six subjects per cell might be successful. An extension of the second experiment might be equally effective, but it is thought that the Advertising Standards Authority might not find this acceptable. The consultant also points out that statistical significance would have been achieved in the second experiment if the residual mean square had been smaller. He wonders if the variability from person to person could have been reduced by exercising tighter control over the hair washing and/or the assessment of dandruff. A lengthy discussion of the practical difficulties and possible alternative procedures comes to fruition when the consultant points out the advantages of asking each participant to use both shampoos. This could be achieved in two ways; by using a cross-over design or by using a split-plot approach. It is decided to use the latter and the third experiment is planned as follows.

Each subject will attend the company hair salon once a week for four weeks. At the first visit two dandruff scores will be given, one for the left side of the head and one for the right side. After dandruff assessment, one side of the head will be washed using Cleartop and the other side using Softsheen. Half of the subjects would have Cleartop on the left and the other half would have Softsheen on the left. This would be repeated exactly at the second and third visits so that subject X would have Cleartop on the same side at each wash. On the fourth visit dandruff assessment will be carried out but no washing will take place.

Obviously this procedure will cost much more per subject than the procedures used in the earlier experiments. On the other hand, the number of subjects required should be much less for two reasons:

1. Each subject yields twice as many scores at each assessment.
2. The comparison of the two shampoos is carried out with greater precision because it is no longer contaminated by person-to-person variation.

The statistical consultant suggests that the half-head design could be more powerful using only eight subjects than the second experiment using 32 subjects. However, he is unable to quantify the increased precision as the earlier experiments have not yielded any estimate of 'within person variability'. Gordon Staples being fearful of yet another 'very interesting but not quite significant' conclusion, decides to make available sufficient funding for sixteen participants. Thus four subjects are recruited for each of the four categories: dry–short hair, dry–long hair, greasy–short hair and greasy–long hair. The decrease in dandruff from the first to the fourth visit is calculated for each side of each subject. The results are given in Table 3.

Q1 Table 1 and Table 3 look very similar. This might seem to imply that the two experiments would be equally effective. However, the similarity of the two data tables is misleading, for there are important

Table 3
Decrease in dandruff: third experiment

	TYPE OF HAIR														
	DRY								GREASY						
	LENGTH OF HAIR								LENGTH OF HAIR						
SHAMPOO	SHORT				LONG				SHORT				LONG			
Cleartop	2	0	3	1	3	3	2	0	3	3	4	1	2	4	4	0
Softsheen	1	1	3	−1	4	3	0	1	0	1	1	−1	1	1	1	−2

differences between the two experiments. (See, Cochran and Cox, 1957, p. 293; Johnson and Leone, 1964, p. 227; Cox and Snell, 1981, p. 135; Mead and Curnow, 1983, p. 96.)

(a) Explain why the experiment three is more likely than experiment two to reveal a difference between the two shampoos, if a difference exists.

(b) Is experiment three more powerful than experiment two in its abililly to detect an interaction between shampoo and hair type?

(c) Is experiment three more powerful than experiment two in its ability to detect an interaction between hair type and hair length?

Q2   Analysis of variance on the data in Table 3 will yield a 'between subjects residual mean square' and a 'within subjects residual mean square'. Roughly assess the magnitude of either or both of these mean squares using evidence available from the first two experiments.

Q3   Carry out an analysis of variance on the data in Table 3. The references given in Question 1 offer some guidance on the analysis of data from a split plot experiment. However, you may find that you need access to a suitable computer program. The following procedure is not the most efficient but it is more meaningful than many alternatives:

(a) Carry out a one-way analysis of variance to break down the total variation into two components: 'between subjects' and 'within subjects'.

(b) Break down the 'between subjects sum of squares' into three components, each having one degree of freedom, plus a between subjects residual.

(c) Break down the 'within subjects sum of squares' into four components, each having one degree of freedom, plus a within subjects residual. (Perhaps this is most easily achieved using the within person differences.)

Q4   Having completed the analysis of variance in Question 3(c), what conclusions can you draw concerning the effectiveness of the two shampoos with different types and lengths of hair?

FURTHER INVESTIGATION

Gordon Staples is very pleased with the outcome of the third experiment. His initiative, bringing in the statistical consultant, has been completely vindicated. In fact, it now seems remarkable that he felt such doubts at the time. He notes that the half-head design was just as powerful as the statistician had predicted and he suspects that eight subjects might have been sufficient after all. However, he decides not to mention this point in his report.

The evidence obtained from the third experiment will be used to support a major advertising campaign. This will be executed by an advertising agency which has been used in the past, and will take several months to organize. However, Gordon Staples has plans for an immediate, though much smaller, project.

You will recall that Gordon Staples is a Marketing Manager with Beauticare Limited. Though the main business interest of Beauticare lies in cosmetics and toiletries, it does have various subsidiaries which are only loosely connected with this area. Amongst the subsidiaries are four department stores. In the past Gordon has made use of these stores to assess the value of various marketing initiatives. He plans to do so again in order to explore the reaction of customers to different ways of presenting his advertising claim.

A short video presentation is produced in which Cleartop and Softsheen are explicitly compared. Four versions of the video tape are made, with identical pictures but different sound tracks. In version X the verbal message concentrates upon the superiority of Cleartop, whilst in version Y the message emphasizes the inferiority of Softsheen. Both X and Y messages are recorded using a male voice and a female voice. Gordon wishes to assess which of the four videos will prove superior when played in the department stores. He is aware that all four stores use video presentations in various locations, including the restaurant, the accounts area and the sales floors. He resolves to position one video player at the entrance to the store and the other player within the cosmetic department, with a Cleartop sales counter next to each. After considering the practical difficulties associated with these arrangements Gordon produces the designs in Tables 4 and 5.

Table 4
Gordon Staples' first design

AUDIO MESSAGE	SEX OF VOICE	LOCATION	STORE
X	Male	Store entrance	Brudfax
Y	Male	Store entrance	Shoremount
X	Male	Cosmetic department	Brudfax
Y	Male	Cosmetic department	Shoremount
X	Female	Store entrance	Shoremount
Y	Female	Store entrance	Brudfax
X	Female	Cosmetic department	Shoremount
Y	Female	Cosmetic department	Brudfax

Table 5
Gordon Staples' second design

AUDIO MESSAGE	SEX OF VOICE	LOCATION	STORE
X	Male	Store entrance	Brudfax
Y	Male	Store entrance	Shoremount
X	Male	Cosmetic department	Woodhill
Y	Male	Cosmetic department	Cragvale
X	Female	Store entrance	Brudfax
Y	Female	Store entrance	Shoremount
X	Female	Cosmetic department	Woodhill
Y	Female	Cosmetic department	Cragvale

Gordon Staples is unable to decide which of the two experiments is likely to tell him more about the four versions of the videos and the two locations. Furthermore, he suspects that there might be an experimental design which would be superior to both. He decides, therefore, to seek further advice from the statistical consultant who helped with the shampoo comparison experiment.

The statistician asks many searching questions about the objectives of the experiment and the difficulties associated with the two proposed designs. He learns that Gordon Staples intends to keep the video players in the specified locations for exactly one week. A shorter period could lead to biased results as there is considerable day-to-day variation in any department store. On the other hand, he would not wish the whole experiment to last longer than two weeks as this would increase the likelihood of customers being influenced by other advertising or promotions. The statistician suggests it is impossible to carry out, in one week, an experiment that will achieve all of Gordon Staples' objectives. However, he offers the two designs in Tables 6 and 7, which he claims are superior to those in Tables 4 and 5.

Table 6
The statistician's first design

AUDIO MESSAGE	SEX OF VOICE	LOCATION	STORE
X	Male	Store entrance	Brudfax
Y	Male	Store entrance	Shoremount
X	Male	Cosmetic department	Shoremount
Y	Male	Cosmetic department	Brudfax
X	Female	Store entrance	Shoremount
Y	Female	Store entrance	Brudfax
X	Female	Cosmetic department	Brudfax
Y	Female	Cosmetic department	Shoremount

Table 7
The statistician's second design

AUDIO MESSAGE	SEX OF VOICE	LOCATION	STORE
X	Male	Store entrance	Brudfax
Y	Male	Store entrance	Shoremount
X	Male	Cosmetic department	Woodhill
Y	Male	Cosmetic department	Cragvale
X	Female	Store entrance	Cragvale
Y	Female	Store entrance	Woodhill
X	Female	Cosmetic department	Shoremount
Y	Female	Cosmetic department	Brudfax

The following questions require you to compare the various experimental designs above.

Q5   The statistician has asserted that his design in Table 6 is superior to Gordon Staples' design in Table 4. Do you agree? If so, why? If not, why not?

Q6   In what respects is the design in Table 7 superior to the design in Table 5? (There are many ways in which the two experiments can be compared. You might wish to draw up a design matrix for each and examine the intercorrelations.)

Q7   Compare the two experimental designs in Tables 6 and 7. What are the advantages and disadvantages of each? (See Cochran and Cox, 1957, Ch. 6; Johnson and Leone, 1964, Ch. 15; Daniel, 1976, Ch. 10; Box, Hunter and Hunter, 1978, Ch. 10; Caulcutt, 1983, Ch. 13.)

REFERENCES

Box, G. E. P., Hunter, W. G., and Hunter, J. S. (1978). *Statistics for Experimenters*, Wiley, New York.

Caulcutt, R. (1983). *Statistics in Research and Development*, Chapman and Hall, London.

Cochran, W. G., and Cox, G. M. (1957). *Experimental Designs*, Wiley, New York.

Cox, D. R., and Snell, E. J. (1981). *Applied Statistics: Principles and Examples*, Chapman and Hall, London.

Daniel, C. (1976). *Applications of Statistics to Industrial Experimentation*, Wiley, New York.

Johnson, N. L., and Leone, F. C. (1964). *Statistics and Experimental Design*, Wiley, New York.

Mead, R., and Curnow, R. N. (1983). *Statistical Methods in Agriculture and Experimental Biology*, Chapman and Hall, London.

SUPPLEMENTARY QUESTIONS

1. The split plot experiment gets its name from its agricultural origins, where plots of land can be split into sub-plots. In this assignment we have referred to human heads as plots and half-heads as sub-plots. Obviously the split plot design can be applied to a variety of situations and it adapts well to a two-stage production process.

   Four batches of intermediate are made at the first stage using a 2^2 factorial experiment to explore the effects of temperature and feedrate. Each batch of intermediate is split into four equal parts to feed a total of 16 runs at the second stage where four replicates of a 2^2 factorial experiment are used to explore the effects of belt speed and thickness in the production process. The inherent variability of the second stage is less than that of the first stage. Which main effects and two-factor interactions would you compare with the second-stage residual and which would you compare with the first-stage residual?

2. Comment on the assertion that 'blocking gives us a more sensitive comparison of treatments by reducing the residual variance. However, in many situations the residual could be reduced more easily by paying attention to detail during the conduct of the experiment and the measurement of results'.

3. What assumptions underly the significance tests you carried out in the first part of this assignment? How would you check that the assumptions were not violated? How could you modify your data analysis if you found an assumption was violated?

Assignments in Applied Statistics
Edited by S. Conrad

A Hill Climbing Approach to Find the Optimum Conditions for Satisfying Quality and Quantity Specifications

Roland Caulcutt

Management Centre, University of Bradford

OUTLINE

In agricultural research it may not be possible to carry out more than one experiment per year on any plot of land. Thus the researcher must try to learn as much as possible from the experiment, as it will be twelve months before there is another opportunity. Many texts, which have been strongly influenced by the agricultural tradition, give the impression that success is achieved by carrying out one large experiment which will reveal all. However, in other research areas, many scientists prefer to carry out a sequence of small experiments. This is particularly true in the development of industrial processes. In this assignment we will examine an attempt to improve the performance of a simple drying process in order to meet a tightened specification. The scientist in charge of the investigation adopts a well-established procedure based on fractional 2^n factorial experiments followed by a 3^n factorial as the optimum is approached. Progress is not as smooth as you might expect.

KEYWORDS

sequential experiments, steepest ascent, fractional factorial experiments at two levels, factorial experiments at three levels, optimum conditions, multiple regression analysis

Chemspec plc makes speciality chemicals which are used in the manufacture of synthetic rubber and flexible adhesives. Customers for Chemspec products include the major tyre manufacturers, who exert increasing pressure on their suppliers to attain very high quality. In doing so the tyre manufacturers are simply responding to the conditions imposed upon them by their major customer, the Ford Motor Company. It is now clear that the quality campaign initiated by Ford in 1982 has resulted in the spread of supplier auditing backwards through a great diversity of manufacturing companies.

The effect of this pressure over a number of years has been to raise the quality of certain Chemspec products and to motivate the workforce to achieve this high quality with consistency. However, it would be unwise to assume that the customer will be entirely satisfied with any product for very long as continuous improvement is now the norm throughout the industry.

Dr Rathbone is the Production Manager at Chemspec. He is currently very concerned that two tyre manufacturers have both recently requested changes in the specification for certain rubber chemicals. He suspects that one particular change, the reduction of moisture content, might be difficult to achieve. The chemicals leave the reaction vessel in a paste form, which is reduced to powder in a continuous dryer. Dr Rathbone is aware that the dryer is already overloaded and any attempt to further reduce moisture content of the powder might create a severe bottleneck.

The dryer is basically a continuous belt moving over a heated surface. The paste is extruded onto the moving belt and the heat from below causes moisture in the paste to evaporate. This evaporation is assisted by blowing hot air over the surface of the paste. The performance of the dryer can be influenced by adjustment of:

(a) the belt speed,
(b) the heater temperature,
(c) the air temperature,
(d) the air fan speed,
(e) the paste thickness.

The paste extruded onto the belt has a moisture content of about 30 per cent. The powder coming off at the end of the belt must have a moisture content of less than 1.0 per cent if it is to meet the new specification. This will not be achieved if the belt speed is too high and/or the paste thickness is too great. Obviously it would be possible to get increased throughput if the moisture specification were ignored. On the other hand, it would be easy to achieve a low moisture content if the operator was not under pressure to achieve a high throughput.

Dr Rathbone initiates an investigation of dryer performance which proceeds in four stages. At stage 1 the effect of all five independent

Table 1
The first experiment

RUN	BELT SPEED (cm/s)	HEATER TEMPERATURE (°C)	AIR TEMPERATURE (°C)	FAN SPEED (r.p.m.)	PASTE THICKNESS (cm)	MOISTURE CONTENT (%)	THROUGHPUT (kg/h)
4	3.5	74	30	600	4.8	3.15	198
7	3.5	74	35	650	5.2	3.63	217
1	3.5	78	30	650	5.2	3.45	220
8	3.5	78	35	600	4.8	2.92	205
5	3.7	74	30	650	4.8	4.25	216
3	3.7	74	35	600	5.2	5.11	228
2	3.7	78	30	600	5.2	4.95	235
6	3.7	78	35	650	4.8	4.02	210

Table 2
The second experiment

RUN	BELT SPEED (cm/s)	HEATER TEMPERATURE (°C)	AIR TEMPERATURE (°C)	FAN SPEED (r.p.m.)	PASTE THICKNESS (cm)
9	3.4	79	36	650	4.7
10	3.3	80	36	650	4.6
11	3.2	80	37	650	4.5
12	3.1	80	37	650	4.4

variables is explored using the experiment in Table 1. The design of this first experiment is based on a similar experiment described in Davies (1978, p. 512). The eight experimental runs are carried out in random order, each lasting for approximately one hour.

The data in Table 1 is analysed by John Cameron, an R & D chemist who has studied statistics on several short courses but has little practical experience. He attempts to assess the changes in operating conditions that will lead to a reduction in moisture content. His aim is to travel down the 'moisture slope' following the line of steepest descent, using those popular guide books Box, Hunter and Hunter (1978, Ch. 15) and Davies (1978, Ch. 11). His analysis of the data leads him to the conclusion that the four runs specified in Table 2 will give progress in the right direction.

John Cameron is confident that the four runs specified in Table 2 will give lower moisture content than the eight runs in Table 1. He also expects that runs 9, 10, 11 and 12 will give reduced throughput, since he has asked for reduced belt speed and paste thickness. His analysis of the data from the first experiment showed that throughput was dependent on belt speed and paste thickness. However, it was realised before the investigation began that throughput would be determined by these two independent variables alone and would not be related to the other three. Thus it should be possible to reduce moisture content without reducing throughput simply by increasing the heater temperature, the air temperature and the fan speed whilst keeping belt speed and paste thickness constant. However, the upper limit of fan speed (650 r.p.m.) has now been reached and it would be unsafe to raise the heater temperature above 80°C. John Cameron fears, therefore, that some reduction in throughput will be unavoidable if moisture content is to be brought down to within the new specification. He should get some indication of the size of the problem when the results of the second experiment are available.

The results in Table 3 confirm John Cameron's predictions. From run 9 to 10 to 11 to 12, there is a reduction in moisture content. If this line of steepest descent were followed for a further two steps he would be very

267

Table 3
The second experiment

RUN	BELT SPEED (cm/s)	HEATER TEMPERATURE (°C)	AIR TEMPERATURE (°C)	FAN SPEED (r.p.m.)	PASTE THICKNESS (cm)	MOISTURE CONTENT (%)	THROUGHPUT (kg/h)
9	3.4	79	36	650	4.7	2.45	190
10	3.3	80	36	650	4.6	2.23	186
11	3.2	80	37	650	4.5	1.57	169
12	3.1	80	37	650	4.4	1.49	172

Table 4
The third experiment

RUN	BELT SPEED (cm/s)	HEATER TEMPERATURE (°C)	AIR TEMPERATURE (°C)	FAN SPEED (r.p.m.)	PASTE THICKNESS (cm)	MOISTURE CONTENT (%)	THROUGHPUT (kg/h)
14	3.0	80	38	650	4.0	1.23	135
19	3.0	80	38	650	4.4	1.39	163
13	3.0	80	40	650	4.0	1.04	149
16	3.0	80	40	650	4.4	1.44	155
18	3.2	80	38	650	4.0	1.47	155
15	3.2	80	38	650	4.4	1.81	174
20	3.2	80	40	650	4.0	1.59	152
17	3.2	80	40	650	4.4	1.68	159

close to the 1 per cent moisture content demanded by the new specification. Unfortunately the reduction in moisture content is accompanied by a reduction in throughput.

Having successfully followed, in the second experiment, the path of steepest descent predicted by the first, John Cameron now wishes to consolidate this progress. To obtain more reliable evidence than that provided by runs 9 to 12 he carries out a further factorial experiment.

John Cameron is a little disappointed when he sees the moisture contents in Table 4. He had expected the third experiment to take him well into the region where moisture content would be less than 1 per cent. It appears that he has only just reached the edge. Furthermore, he has had to pay a high price for his limited success, since the throughput is now considerably lower than in the first experiment. His analysis of the data in Table 4 suggests a path of steepest descent which differs little from the line he has followed so far. He decides to try three more runs in this direction. He is almost certain that these will give the required moisture content, but he is equally confident that they will also give a further reduction in throughput. In an attempt to recover a little of the lost throughput without any increase in moisture content he proposes four additional runs on a line perpendicular to the path of steepest descent. These are runs 24 to 27 in Table 5.

Having completed the fourth experiment John Cameron is not sure what conclusions he can safely draw from all the data he has accumulated. Must he base his conclusions on the last seven runs, or could he put together the data from all four experiments? The texts he has followed, Box, Hunter and Hunter (1978) and Davies (1978), both seem to imply that he should have carried out a three-level factorial experiment rather than exploring along a line perpendicular to the path of steepest descent. John decides that he will visit the company statistician.

Q1 What type of experimental design was used for the first experiment?

Q2 From the results of the first experiment it is possible to calculate the mean response and seven other estimates. Determine the defining contrasts and list the other effects in seven alias groups. (See Cochran and Cox, 1957, p. 244; Cox, 1958, p. 248; Johnson and Leone, 1964, pp. 206 and 248; Davies, 1978, p. 451; Caulcutt, 1983, p. 208.)

Q3 Use the results of the first experiment to determine the path of steepest descent for the reduction of moisture content. (See Cochran and Cox, 1957, p. 359; Box, Hunter and Hunter, 1978, p. 517; Davies, 1978, p. 513.)

Q4 In the light of the alias groups in Question 2, suggest how the path of steepest descent could be misleading.

Table 5
The fourth experiment

RUN	BELT SPEED (cm/s)	HEATER TEMPERATURE (°C)	AIR TEMPERATURE (°C)	FAN SPEED (r.p.m.)	PASTE THICKNESS (cm)	MOISTURE CONTENT (%)	THROUGHPUT (kg/h)
21	2.9	80	40	650	3.8	0.82	126
22	2.8	80	40	650	3.6	0.68	117
23	2.7	80	40	650	3.4	0.53	116
24	3.4	80	40	650	3.2	1.58	139
25	3.2	80	40	650	3.6	1.20	134
26	2.8	80	40	650	4.4	0.75	156
27	2.6	80	40	650	4.8	0.63	145

Q5   John Cameron is happy to assume that throughput (y) is dependent only on the belt speed (x) and the paste thickness (z). Furthermore, he is willing to assume that $y = axz$. Use the data from the first experiment to estimate a and then produce a contour diagram to illustrate the dependence of y on x and z.

Q6   Use the values of belt speed and paste thickness from Tables 1, 3, 4 and 5 to plot 27 points on your contour diagram. Write next to each point the moisture content for that run.

Q7   Use the diagram from Question 6 to assess the best belt speed and paste thickness to use in order to achieve maximum throughput, whilst satisfying the following customers:
     (a) Customer A specifies that the moisture content should not exceed 1 per cent.
     (b) Customer B specifies that the moisture content should not exceed 2 per cent.

Q8   What throughput do you think will be achieved using the belt speeds and paste thickness that you recommended in Question 7?

Q9   John Cameron is prepared to carry out four additional experimental runs to help you improve upon the recommendations you made in Question 7. What operating conditions would you wish him to use for these four runs?

Q10 In this assignment you have not carried out any significance tests. How, then, can you be confident that the 'line of steepest descent' followed by John Cameron is not simply chasing random variation?

FURTHER INVESTIGATION

John Cameron is very satisfied with the four experiments he has carried out so far. He is particularly pleased with runs 24 to 27, which were based on his own intuition rather than on the advice offered by his 'guide books'. However, he believes that a change of approach may now be needed, because he is close to the optimum conditions that he is seeking. Both Box, Hunter and Hunter (1978) and Davies (1978) recommended the use of three-level factorial designs in the vicinity of an optimum.

In what John Cameron hopes will be the final experiment (see Table 6) the heater temperature is maintained at 80°C, the air temperature at 40°C and the fan speed at 650 r.p.m. No further increase is possible with any of these three variables. The paste thickness will have three levels, 4.6, 4.8 and 5.0 centimetres. The belt speed will also have three levels, 2.7, 2.8 and 2.9 centimetres per second.

Table 6
The fifth experiment

RUN	BELT SPEED (cm/s)	PASTE THICKNESS (cm)	MOISTURE CONTENT (%)	THROUGHPUT (kg/h)
28	4.8	2.8	0.99	151
29	4.6	2.8	0.86	155
30	4.6	2.9	1.13	166
31	5.0	2.8	0.95	175
32	4.8	2.7	0.80	150
33	5.0	2.9	1.25	172
34	4.6	2.7	0.84	138
35	5.0	2.7	0.84	170
36	4.8	2.9	0.92	173

The analysis of the results of the fifth experiment take John Cameron into what, for him, are uncharted waters. In the previous experiments, with each variable at only two levels, he was happy to follow the methods of analysis recommended in the statistical texts. These methods seemed to be compatible with common sense. However, he finds that many texts do not offer detailed advice on the analysis of results from 3×3 factorial experiments. Furthermore, those which do cover three-level designs are very difficult for someone with little mathematical foundation. After spending much time searching for something simpler, he decides to use the method of analysis given in Davies (1978, Ch. 8).

John Cameron's analysis of the data in Table 6 leads him to some rather unexpected conclusions. He had been led to believe that the 3×3 factorial experiment would enable him to estimate the parameters of a quadratic response surface, which would indicate the operating conditions that would give minimum moisture content. However, his analysis appears to reveal that moisture content depends only upon the paste thickness and that this relationship is linear. Thus, by implication, moisture content does not depend upon the belt speed.

These conclusions not only conflict with his earlier findings but they also imply that John Cameron was wrong to expect the optimum conditions to lie within the experimental region covered by the fifth experiment. He is very disappointed with the outcome of this experiment, but he wonders if the company statistician might be able to suggest a 'better' method of analysis.

Q11   John Cameron used analysis of variance to analyse the data in Table 6. He took account of the linear effects and the quadratic effects of both independent variables. He used a further degree of freedom for the linear interaction, leaving only three degrees of freedom for the residual. Carry out a similar analysis using the moisture results in Table 6. Do you agree with John's conclusions? (See Davies, 1978, Ch. 8.)

Q12   Can you suggest why the fifth experiment has not led to such clear conclusions as did the first four experiments?

Q13   The seven runs in Table 5 and nine runs in Table 6 were all carried out using a heater temperature of 80°C, an air temperature of 40°C and a fan speed of 650 r.p.m. Combine the data from all sixteen runs and use multiple regression analysis to explore the response surface. What conclusions can you draw concerning the belt speed and paste thickness required to achieve less than 1 per cent moisture?

Q14   Can you suggest any reason why it might be unwise to combine the data from the two experiments? How could you make allowance, in your regression analysis, for the fact that the data came from two separate experiments?

Q15   The eight runs in Table 4 were also carried out using a heater temperature of 80°C and a fan speed of 650 r.p.m. However, they differ from the fifteen runs in Tables 5 and 6 because two air temperatures were used. For runs 14, 15, 18 and 19 subtract 0.02 from the moisture content to allow for the lower air temperature; then combine these data with that from Tables 5 and 6. Carry out a second regression analysis to explore the response surface. Do you wish to revise the conclusions you drew in Question 13?

REFERENCES

Box, G. E. P., Hunter, W. G., and Hunter, J. S. (1978). *Statistics for Experimenters*, Wiley, New York.

Caulcutt, R. (1983). *Statistics in Research and Development*, Chapman and Hall, London.

Cochran, W. G., and Cox, G. M. (1957). *Experimental Designs*, Wiley, New York.

Cox, D. R. (1958). *Planning of Experiments*, Wiley, New York.

Davies, O. L. (1978). *The Design and Analysis of Industrial Experiments*, Longman, London.

Johnson, N. L., and Leone, F. C. (1964). *Statistics and Experimental Design, Volume 2*, Wiley, New York.

SUPPLEMENTARY QUESTIONS

1. How does the hill climbing approach used in this assignment differ from the technique of evolutionary operation (EVOP)? (See Box, Hunter and Hunter, 1978.)

2. As we approach the optimum conditions the 2^n factorial experiment must be abandoned in favour of a design that allows us to estimate quadratic effects. In this assignment we switched to a 3^n factorial design, but there are alternatives. For example, the 2^n factorial experiment could be extended to give an orthogonal composite design. What are the advantages and disadvantages of the composite design compared with the 3^n factorial in this situation? (See Box, Hunter and Hunter, 1978; Davies 1978, p. 232.)